BRITISH AND JAPANESE MILITARY LEADERSHIP IN THE FAR EASTERN WAR, 1941–1945

CASS SERIES: MILITARY HISTORY AND POLICY
Series Editors: John Gooch and Brian Holden Reid
ISSN: 1465–8488

This series will publish studies on historical and contemporary aspects of land power, spanning the period from the eighteenth century to the present day, and will include national, international and comparative studies. From time to time, the series will publish edited collections of essays and 'classics'.

BRITISH AND JAPANESE MILITARY LEADERSHIP IN THE FAR EASTERN WAR, 1941–1945

Edited by Brian Bond
and Kyoichi Tachikawa

Routledge
Taylor & Francis Group

LONDON AND NEW YORK

First published 2004
by Frank Cass

This edition published 2012 by Routledge

2 Park Square, Milton Park, Abingdon, Oxon OX14 4RN

711 Third Avenue, New York, NY 10017

Routledge is an imprint of the Taylor & Francis Group, an informa business

Transferred to Digital Printing 2006
First issued in paperback 2012

Typeset in Garamond by
Keystroke, Jacaranda Lodge, Wolverhampton

British Library Cataloguing in Publication Data
A catalogue record for this book is available from the British Library

Library of Congress Cataloging in Publication Data
British and Japanese military leadership in the Far Eastern War,
1941–45 / editors, Brian Bond and Kyoichi Tachikawa.—1st ed.
p. cm. — (Cass series—military history and policy ; no. 17)
Includes bibliographical references and index.
1. World War, 1939–1945—Campaigns—Malay Peninsula.
2. World War, 1939–1945—Campaigns—Burma.
3. Strategy—History—20th century 4. Command of
troops—History—20th century. 5. Leadership—Japan—
History—20th century. 6. Leadership—Great Britain—History—
20th century. I. Bond, Brian.   II. Tachikawa, Kyåoichi, 1966–
III. Title.   IV. Series.

D767.5.B67 2004
940.54′2591—dc22   2004004294

ISBN13: 978-0-714-65659-5 (hbk)
ISBN 13: 978-0-415-64622-2 (pbk)

CONTENTS

CONTENTS

CONTRIBUTORS

Kanji Akagi is Professor, Keio University. He received an MA from Keio University in 1980 and a PhD in 1989. He is author of (in Japanese) *The Origins of the Vietnam War* and (in Japanese) *Policy and Strategy in the Second World War*.

Kenichi Arakawa is Associate Professor at the National Defense Academy (NDA), Japan, and a Colonel (retired), JGSDF. He was awarded an MA from Toyo Eiwa University in 1997 and is a PhD candidate at Hitotsubashi University. He is author of (co-authored, in Japanese) *New Perspectives on the Sino-Japanese War* and (co-authored, in Japanese) *The Manchurian Incident Revisited After Seventy Years*.

Brian Bond is Emeritus Professor of Military History at King's College London and President of the British Commission for Military History. *Festschrift* to mark his retirement in 2002, *The British General Staff*, edited by David French and Brian Holden Reid, was published by Frank Cass and awarded the Templer Prize Medal for Military History. He is author of many books on military history, the most recent being *The Unquiet Western Front: Britain's Role in Literature and History* (2002).

Carl Bridge is Professor, King's College, London and head of the Menzies Centre for Australian Studies there. He is author of *Munich to Vietnam: Australia's Relations with Britain and the United States since the 1930s* and *Between Empire and Nation: Australia's External Relations from Federation to the Second World War*.

Michael Dockrill is Emeritus and Visiting Professor of Diplomatic History, King's College, London. He received his PhD from the London School of Economics in 1969 and is author or editor of many books on British foreign policy and stratgy, the most recent being *British Establishment Perspectives on France 1936–40* and (edited, with Philippe Chassaigne) *Anglo-French Relations 1898–1998*.

Saki Dockrill is Professor of Contemporary History and International Security, King's College, London. She is a Fellow of the Royal Historical Society and has written extensively on defence and security relations in Europe and Asia in the twentieth century; her publications include *Britain's Policy for West German Rearmament*; (as editor) *From Pearl Harbor to Hiroshima: The Second World War in Asia and the Pacific, 1941–1945*; *Eisenhower's New Look Security Policy* and, most recently, *Britain's Retreat From East of Suez: The Choice Between Europe and the World?*

Graham Dunlop is a Colonel (retired), Royal Marines. He is currently completing a PhD thesis at the University of Edinburgh. He served in Northern Ireland, the Mediterranean, the USA, Hong Kong, Iraq and Bosnia, before retiring in 1997.

Kyoichi Tachikawa is Senior Research Fellow, Military History Department, National Institute for Defense Studies, Tokyo, Japan. He was awarded an MA from Sophia University in 1990 and a PhD in 1999. He is author of (in Japanese) *World War II and French Indochina: A Study of French–Japanese Collaboration*.

Ryoichi Tobe is Professor at the National Defense Academy (NDA), Japan. He received an MA from Kyoto University in 1973 and a PhD in 1991. He is author of (in Japanese) *Japanese Peace-Feelers in the China Incident* and *Army of Paradox*.

Hisayuki Yokoyama is Associate Professor at the National Defense Academy (NDA), Japan, and a Lieutenant-Colonel in the JASDF. He was awarded an MA from Obirin University in 2003.

SERIES EDITOR'S PREFACE

The subject of military leadership is at once one of the most fascinating and yet intractable in the canon of war studies. The precise qualities demanded by leaders at all levels are perhaps easy to enumerate in list form but more difficult to nurture, because they depend not only on personal qualities but on environment, culture, opportunity, and above all on the capacity to learn from past experience. The perspective in this volume is comparative, and all students of the Burma Campaign will learn a lot from it.

The Burma Campaign is characterised by dazzling Japanese successes at the beginning, and an equally dazzling British riposte at the end. A long period of stalemate occurred in the middle, as each side sized the other up and simultaneously attempted to come to terms with the grave problems posed by climate and terrain. In this crucial phase the balance of success shifted from the initial victors to the defeated. Japanese commanders revealed themselves tactically bold but strategically irresolute. Consequently, they drew far too heavily on the warrior ethos of their troops. They neglected logistics, medical treatment, and left casualties to die, and tended to assume good morale would continue rather than attempt to nourish it. Even the bravest soldiers will not continue to fight if the very basics of their daily needs are neglected.

The British problem was the converse. At first they stumbled tactically, but more rapidly solved their strategic problems. Defeat had impressed on them the enormous significance of logistics, communications and rapid medical treatment in such an inhospitable environment. Leadership consists of far more than adopting courageous poses. The British commander, Slim, developed a style of leadership well suited to the arduous conditions. His greatest achievement was binding together a disparate coalition of different races, cultures and military styles, and instilling it with the will to win.

These fine, meticulous essays also offer a warning of the dangers of accepting uncritically crude racial stereotypes of the enemy. First, the British thought the Japanese incompetent, then after the spring of 1942 jumped to the opposite extreme of believing them supermen. The Japanese thought the British 'gin and tonic' warriors, addicted to the tennis courts and mess dinners. All of these caricatures were wrong. Let us hope that the editors of this scholarly collection are successful in encouraging less chauvinistic, one-sided approaches to the study of this perennially fascinating campaign.

Brian Holden Reid
Series Co-editor

FOREWORD

It is my firm conviction that by improving Japan's standards of using the 'comparative method' of history to that of the international level, it can contribute to the mutual understanding of war history. Thus, the first International Forum on War History (IFWH), 'War Leadership: Britain and Japan in the Second World War', was organized in 2000.

War is the largest-scale military action and reaction that a state can use in order to force its intentions unilaterally on an opponent, and history is often abused by victorious sides. This is one of the reasons why each party at war has a different perception of a particular war, sometimes harming the relations among the states. Some 60 years after the Pacific War, Japan faces serious historical problems over the interpretation of the war with its former adversaries in the Asian-Pacific region, including China and South Korea. The purpose of the IFWH was not only to learn valuable historical lessons, but also to help build better and more constructive relations with the states involved.

At the IFWH, we have employed the comparative method to encourage a better and deeper understanding of war history. It is to be hoped that this method will ensure objectivity, thus enhancing authenticity. At the same time, we have to be careful not to rely too much on this particular method. Misunderstanding and lack of understanding, or perhaps an unwillingness to compromise caused by cultural differences, may result not only in futile discussions but also render the comparisons fruitless.

As Professor Brian Bond aptly states, 'Failures and errors of judgement are easy to list with the wisdom of hindsight, but learning 'lessons' which can be applied in different circumstances is much harder.' This is a sound and thought-provoking comment, suggesting realistic but profound implications.

Professor Bond expressed his interest in the IFWH on the occasion of the 26th International Congress of Military History in August 2000, and suggested how to organize future conferences. I appreciate his thoughtful suggestions. We have to also thank Brian for finding the speakers from the UK and Australia. I would like to express my gratitude to these distinguished scholars: Professor Michael Dockrill, Professor Saki Dockrill, Col. Graham Dunlop, and Professor Carl Bridge. They made excellent presentations at the IFWH.

At the first IFWH, keynote speaker, Professor K. Itoh, President of the Japan Forum on International Relations, gave us instructive suggestions about how we could think about a 'new war' seriously. Two of the discussants, Professor T. Tanaka, of Hitotsubashi University, and Professor Emeritus K. Miwa, of Sophia University, have made important contributions to stimulate debate. I appreciate the deep insight of these three distinguished scholars.

Of the Japanese speakers, I was honoured that Professor R. Tobe, of the National Defense Academy, and Professor K. Akagi, of Keio-Gijuku University, accepted our invitation to participate in the IFWH. From its planning stages, they helped our project as supervisors and key members of the preparatory workshop held several times before the IFWH. No doubt, they are among the figures indispensable for the future development of the IFWH.

The former Presidents of NIDS, Mr Yanagisawa and Mr Shinkai, in particular lent strong support to our project. In addition, all the staff at NIDS worked very hard in hosting the IFWH. I am proud of my colleagues and I would like to express my gratitude for all their efforts.

Prof. and Maj.-Gen. (Rtd JASDF) Yoshinaga Hayashi
Director of Military History Department,
National Institute for Defense Studies,
Defense Agency of Japan

NOTE ON JAPANESE NAMES

Japanese names are shown as family name first followed by given name, except where they are the names of Japanese authors of publications in English, where the Western style of given name first is retained.

INTRODUCTION

Brian Bond

This volume represents the collaborative work of five British-based and five Japanese scholars, who have reappraised the leadership of the two sides in the Malaya and Burma campaigns. In these operations the ebb and flow of military fortunes could scarcely be more dramatically illustrated. In December 1941 and the early months of 1942, Japanese forces overran Malaya, captured Singapore and occupied most of Burma. These remarkable conquests were achieved against heavy numerical odds and with very few casualties. Our critical attention consequently focuses first on the reasons for the Japanese forces' unexpectedly rapid and devastating advances, as compared with the extremely poor performance of British, Indian and Australian units. On both sides, we are primarily concerned with leadership, but clearly other factors such as intelligence, equipment, logistics and morale have to be taken into account.

After a period of stalemate in 1942–43 in Arakan and Assam, the ascendancy gradually shifts to the British and Allied side until, suffering from chronic overstretch stemming from disastrous strategic decisions, the Japanese forces are driven back in appalling conditions from north to south Burma and suffer virtual annihilation.

British historians have, not surprisingly, published copiously on this conversion of 'Defeat into Victory' (the apt title of Slim's memoirs), but recent attention has focused almost obsessively on the exploits of Orde Wingate and the Chindits – who receive only marginal coverage here – whereas we are more interested in how the lessons of 1941–42 were learned and implemented in 1944–45. For their part, the Japanese contributors are able to exploit a wide range of archival sources and to write more frankly than in the past about the failures of leadership, both in Tokyo and in the South-East Asia theatre, which resulted in such an overwhelming defeat in Burma.

In British historiography the surrender of Singapore has understandably been a matter of perennial controversy, some accounts dwelling on pre-invasion warnings which were neglected, others employing hindsight to lambast senior British officers, particularly General Percival. But more general reasons for the abject failure of the defenders in both Malaya and Burma are easy to find.

1

Given Britain's strategic predicament in 1940–41, lucidly explained here by Saki Dockrill, it is not surprising that Malaya and Burma received a low priority for all military needs, including tanks, artillery and, above all, modern aircraft. The small, inexperienced and poorly trained land forces were scattered and ill co-ordinated, mainly because their primary role was seen as the defence of airfields. Morale was brittle, especially among some Indian soldiers already feeling the influence of nationalist propaganda. As with Britain's grand strategy, so also at the theatre level there was no clear or feasible plan to meet a Japanese offensive, particularly a land-based advance down the Malay peninsula. A spirit of complacency was prevalent, and perhaps excusable, but this was accompanied by an underestimation of Japanese martial prowess and daring, which was not. Lastly, as Carl Bridge's contribution shows, British and Australian generalship proved disappointing in failing to check the rapid advance of numerically inferior Japanese forces. As he points out, there were some able commanders at battalion level but their ideas were not adopted in time. Curiously, for British readers, Kyoichi Tachikawa is also quite critical of General Yamashita's style of leadership, despite his successful advance culminating in the capture of Singapore.

Many of the factors mentioned above also applied to Burma. The complete loss of air cover at an early stage was critical, as was also the paucity of tanks, field artillery and modern equipment generally. Slim excelled in commanding Burcorps' epic retreat but he was handicapped by lack of a clear directive about his mission. Was he meant to try to hold a line in southern Burma; protect the oilfields in the centre; or save as much as he could by a precipitate retreat across the Chindwin and towards the sanctuary of the Imphal plain?

In the event, the awesome speed and determination of the Japanese advance left him no choice, and the historical verdict is that Slim did well, not merely to extract the bulk of his exhausted corps, but also to preserve their fighting spirit. This spirit had been sorely tried by inability to cope with Japanese infiltration tactics, which repeatedly set up road blocks in the British rear. Unable at this stage to operate off the roads, British troops were inclined to regard the Japanese infantry as supermen who could operate at will through the jungle. This exaggerated respect, allied to a well-merited fear of the enemy's merciless treatment of wounded and prisoners, and the terrible problems posed by the terrain, climate and disease together presented an enormous challenge to British and Allied leadership at all levels.

Brian Bond, in his reassessment of Slim, shows how these challenges were met at Army level, while Graham Dunlop explores the learning process at corps level and below in a stimulating essay based on his current doctoral research.

Admiral Lord Louis Mountbatten's arrival in the autumn of 1943 to head the newly created South-East Asia Command brought much-needed drive at the top; an improved co-ordination between the three services; and much better relations between the US and Chinese leaders in the theatre. Mountbatten's charisma and prestige also helped to improve the theatre's status in London and Washington but it remained low in the pecking order for such vital needs as transport planes and landing craft. This uncertainty as to whether there would be *any* undertaking to reconquer Burma from north to south makes all the more creditable the training, organising and equipping of Eastern Army (later re-named 'Fourteenth Army' under General Slim) after the disastrous retreat in 1942 and the debacle in Arakan in 1942–43, which led to the dismissal of General Irwin.

Here, we need only note briefly that Slim and his commanders displayed admirable professional skills; first, in assimilating the lessons of defeat and, then, in instilling these in the racially, culturally and linguistically distinctive forces at his disposal, which, in theory, should have been a great handicap in defeating the homogeneous Japanese. In addition to far-reaching improvements in health, medical care and training, Fourteenth Army eventually succeeded in transforming its tactical doctrine and practice. Above all, Slim inculcated a belief that the enemy were not supermen in jungle warfare; their standard tactics of converging hooks through the jungle to impose blocks on rearward communications could be resisted and defeated. This was first demonstrated by the successful defence of the 'admin box' in Arakan. Growing confidence that positions could be held, and isolated groups maintained deep behind enemy lines, entirely by air supply, was strengthened by Wingate's first expedition in 1943.

As regards leadership, Slim was fortunate in the commanders he inherited in 1942, such as Scott and Cowan, but thereafter the selection process was rigorous with the result that by 1944 he led a first-class team at brigade, division and corps levels. On what is now termed the 'mission command' system, Slim encouraged an unusual degree of independent initiative in his senior subordinates, issuing only brief and broad directives. He also recognised that, given the vast distances, formidable terrain and poor communications, logistics and administration were even more important than in a European theatre. Consequently, his chief staff officer was selected for administrative ability rather than for the traditional role of planning operations. There seems to have been a more harmonious and co-operative spirit at Fourteenth Army's headquarters than in its Japanese equivalent; though it must be admitted that a great deal of friction and discord was evident in SEAC and between it and some of the senior US personnel in the theatre.

Graham Dunlop's chapter is particularly revealing on the problems of leadership at battalion level and below. By late 1943, there was clearly a problem with the quality of officers and men being recruited into the British, colonial and Indian Armies. The Indian Army, for example, had expanded from about 190,000 in 1939 to nearly two million by mid-1943. In Burma, junior leaders had not only to demonstrate professional qualities, but also to understand cultural (notably dietary) needs and speak their soldiers' language. This was not always possible in wartime conditions. Indian officers naturally had a closer affinity with their men, but there were simply not enough of them.

Again in contrast to the Japanese experience, Dunlop charts the dramatic improvement in combined-arms tactics in ground combat, and in the co-ordination of air and ground operations by the integration of commanders at corps and army level. At Kohima, for example, it took the combined weight of artillery, air power and tanks to overcome the stubborn Japanese defenders. This chapter also underlines the severe pressure which Slim exerted on his corps commanders, and they in turn on their subordinates. This was espe-cially evident during the siege at Imphal and at the start of the pursuit. The Commander of 2nd Division, Major-General Grover, was dismissed, perhaps unjustly, for insufficient drive during this crisis.

Finally, it is worth noting that, despite remarkable improvements in every sphere of operational skills during the victorious advance to the Irrawaddy, Mandalay and, finally, Rangoon, 'some cracks began to appear in the morale of British troops, caused mainly by war-weariness as well as concerns over affairs at home and repatriation'. One gets the impression that, given these problems and mounting nationalist agitation in India, victory in Burma was achieved in the nick of time.

It remains to consider briefly the assets and defects of Japanese military leader-ship. At the highest level, there was a continual quest for consensus, mainly by means of frequent Imperial Conferences under the nominal but imprecise authority of the Emperor. Senior military and naval commanders had established their dominance in the 1930s by the assassination of unpopular ministers, and this threat was maintained during the war years. As Ryoichi Tobe's essay demonstrates, Tojo was never able to exert overall control of strategic decisions despite his titular occupation of several appointments. Indeed he is depicted as an arch-bureaucrat, a workaholic skilfully manoeuvring to maintain his power base – and posting possible rivals to distant commands – while unable or unwilling to take tough moral decisions as the situation deteriorated.

Curiously the Japanese Army, though not the Navy, reciprocated Western prejudices in despising its European and US enemies, regarding the British, for example, and with some superficial justification, as 'gin-and-tonic warriors' capable only of beating ill-equipped native forces. Their miscalculation regarding the United States' willingness to fight and prodigious manufacturing capacity proved even more serious. This underestimation of the opponents' resolve and staying power partly explains the very modest forces allocated to Southern Army for its widely spread strategic objectives in 1941, and the lack of reserve divisions when they were needed in Burma from 1943 onwards.

At the operational and tactical levels, the Japanese Army had a tremendous initial asset in the prevalence of the warrior spirit, manifested in brutal but effective training, unquestioning obedience of orders, heroic endurance of hardships, a ferocious offensive spirit, an indifference to death and refusal to surrender. Even when the tide of war had turned against them, Japanese soldiers remained tactically adept and dogged in defence: until the closing phase of the war in Burma very few Japanese soldiers were captured alive.

On the debit side, land communications were poor at the start and did not improve. Wheeled and track transportation were greatly inferior to the West's; medical supplies were inadequate; and casualty evacuation was badly neglected. Fundamental weaknesses in the supply system were concealed during the early months of victorious advance against frail opposition, but were cruelly exposed during the long retreat after failure at Imphal. General MacArthur was far from alone in reaching the view that the courageous Japanese infantry had been let down by their commanders in a system where senior appointments were heavily dependent on caste and feudal values. In other words, the Japanese warrior ethos produced not only courage and loyalty but also rigidity, poor staff work and the taking of unjustified risks. As Kenichi Arakawa's chapter demonstrates, the cult of absolute obedience to orders made it extremely diffi-cult for subordinate commanders to query strategic goals which they knew were operationally impossible to achieve, or to halt offensives which were wasting lives and scarce resources to no purpose. With few exceptions, Japanese senior officers evinced considerable physical courage but less moral courage in accepting the necessity for retreat when strategic objectives patently could not be achieved. It is surely not a great exaggeration to conclude that Japan owed its remarkable land conquests essentially to obsolescent warrior traditions, which exploited national values and culture by conducting relentless offensives in a spirit of self-sacrifice with only minimal reliance on up-to-date weapons, communications and supplies. Such a superficially formidable but fragile military machine could only have achieved lasting success against opponents whose martial virtues were as sadly lacking as the Japanese initially assumed. British unpreparedness in 1941 and early 1942 doubtless encouraged these assumptions, but within three years they would prove to be fatally mistaken.

These essays highlight several aspects of Japan's campaigns against Britain and its allies in the Pacific War. The top-level organisation for war, strategic assumptions and initial plans of both sides are described. Styles of generalship and leadership at lower levels receive special emphasis. The enormous problems of communications, supply and soldiers' fitness in some of the worst fighting terrain in the world are also stressed. Two essays reveal sharp contrasts in the employment of air power: the Japanese gave priority to bombing and ground support whereas Britain (and the United States in the Burma–China theatre) put unprecedented reliance on the air transport of troops, supplies and weapons. Not least important, several contributions discuss the moral (and morale) issues which senior commanders on both sides had to deal with, including Japanese reluctance to abandon the Imphal offensive and the British decision to fight on throughout the monsoon months in 1944.

The essays in this book are intended to stimulate reflection and, perhaps, inspire further research on less well-covered aspects of the Pacific War in a new era when past antagonisms can be superseded by collaboration in scholarly endeavours.

1

BRITAIN'S GRAND STRATEGY AND ANGLO-AMERICAN LEADERSHIP IN THE WAR AGAINST JAPAN

Saki Dockrill

Introduction

The war in the Far East – more commonly known as the Pacific War – began with a series of Japanese attacks on Pearl Harbor, the Malayan peninsula, Hong Kong, the Philippines, and islands in the Central Pacific. The Second World War suddenly spread to the other side of Eurasia, spilling into the huge Pacific Ocean. Hearing of Japan's December 1941 onslaught, one official in the British Foreign Office noted in his diary: 'We never thought she would attack us and America at once. She must have gone mad.' In Australia, a newspaper reported it as the 'Gravest hour in the country'. Panic engulfed the United States: people in the West coast claimed that 'Japanese planes had been over San Francisco on the night of 7 December', while in Washington, DC, several senators claimed that they heard that enemy planes were only '150 miles from Washington'. The White House's response was one of relief: 'the indecision was over, and a crisis had come in a way that would unite all our people'.[1] Winston Churchill, the British Prime Minister, reacted similarly: when he retired to bed on the day of Pearl Harbor, he said 'so we had won after all'. The entry of the United States on the side of Britain increased Churchill's confidence in achieving victory in the global conflict. As he recalled, 'Hitler's fate was sealed. Mussolini's fate was sealed. As for the Japanese they would be ground to powder.'[2]

The United States played the leading role in defeating Japan, and, because of this, analysis of Britain's strategy has tended to be overshadowed or neglected by Japanese and other Western writers. Moreover, how Britain developed its Far Eastern strategy prior to, and after, the outbreak of the Pacific War was often misunderstood. It is sometimes argued that Britain greatly under-estimated Japan's military threat and as a result lost its Eastern Empire, or that Churchill 'belittled' the importance of Asia and the Pacific. Alternatively,

6

Britain was already a spent force, and the 'special relationship' forged between Churchill and the US President, Franklin Roosevelt, helped Britain to win victory in the war against Japan.[3] This paper intends to demonstrate that none of these views are entirely correct.

As Britain was a global actor, with vast strategic and trading interests across the world, its grand strategy in war and peace was never formulated in a vacuum, or on a single regional basis. In order to examine Britain's strategy at highest levels during the Pacific War, it is important first to assess where Britain's security interests lay, how they were prioritised during the inter-war period and how they were affected by the outbreak of the Second World War in 1939–41. Secondly, by the time Japan attacked South-East Asia, Britain had formulated a clear strategy about how to prosecute the war on the global level, with the Asian-Pacific war constituting only part of Britain's military undertaking. Although Britain became part of a coalition, with the Anglo-US alliance at the core, this did not in the final analysis compel Britain to change its priorities. How did Britain overcome its initial differences with the United States, and how did the European war impact on the Anglo-US strategy in the Pacific War? Third, the China factor remained a divisive issue in the Anglo-US alliance in their approaches to the war in South-East Asia. The chapter examines how the China factor affected the timing and nature of Britain's war strategy for Burma. As the end of the European war approached, Britain took a serious interest in participating in the final campaign in the war against Japan. The final section discusses Britain's strategy for the defeat of Japan.

Grand strategy

The word 'strategy' is often used in parallel with 'tactics'. According to Carl von Clausewitz, tactics are 'the art of using troops in battle', whereas strategy is the 'art of using battles to win the war'. However, with the passage of time, both words require modification, as they tend to be used in a much wider context than a purely military one, such as 'economic strategy', or 'a strategy for peace'. 'Grand strategy' is more than just strategy in that it embraces both wider goals and long-term goals, that is, the art of managing and controlling national resources to ensure that national interests of all kinds – economic, military, political and cultural (values and beliefs) – are maintained at a minimum cost.[4] National leaders must have the ability to integrate a nation's economic, political and military needs into a grand strategy which will survive not only war but also into the peace. Grand strategy is the art of making both means and ends meet, and of balancing priorities to achieve the nation's requirements.

Britain's grand strategy is no exception, and it was developed in line with the British way in warfare, which was broadly characterised by its preference for economic warfare and for coalition fighting. During the eighteenth and nineteenth centuries, this peripheral strategy allowed Britain to maximise its strengths. Britain deployed its naval superiority to blockade Continental enemies, destroy their fleet and launch raids against the enemy's homeland and overseas possession. Traditionally reluctant to send British troops to the European Continent, Britain was instead able to employ European mercenary troops, rely on the forces of its allies or use imperial forces under Britain's leadership.[5] Britain's diverse worldwide trade and strategic interests meant that Britain needed stability on the European Continent and tried to prevent any single power from becoming so dominant as to upset the balance of power there. After the mid-nineteenth century, Britain preferred more subtle, more indirect and less expensive ways of extending Britain's power and influence, through free trade or treaties, rather than building a formal empire in a closed imperial economic bloc. The so-called 'British Empire' had never been a static or unified entity but it retained an amorphous existence consisting of colonies, protectorates, dependent territories and self-governing dominions. Britain's empire was largely the outcome rather than the end of Britain's external activities, and Britain had never considered its empire as 'defining' its security, strategic or trade interests.[6] The world was then Britain's oyster, and its empire was the manifestation of Britain's predominance.

By the turn of the twentieth century, Britain's fortune began to change, as the spread of the industrial revolution to the Continent and the United States began to erode Britain's supreme naval strength and its wealth. The first break with the traditional grand strategy came during the First World War – while it was still a coalition warfare, Britain had been forced to send large numbers of troops to the Continent, to which Indian and Dominion forces contributed significantly. The end of the First World War brought Britain added responsibilities for enforcing peace in Europe as a result of the defeat of Germany, and also in the Middle East as a result of the collapse of the Ottoman Empire. Altogether, Britain's relative power increased because its rivals either disappeared or became regional powers. The United States had by then surpassed Britain in its economic and industrial strength, but Washington retreated into isolation after the First World War. As a result of all this Britain remained the only global power, even though its economic health had been weakened as a result of the recent war.[7] Britain then came to rely on its empire more for its trade and as the source of manpower in case of Continental and overseas crises.

After the First World War, the Dominions (self-governing states within the British Empire) were more reluctant to sacrifice their manpower to meet the security interests of the mother country, although they expected Britain to protect their security against the growing Japanese threat in the Asian-Pacific theatre. Prior to the outbreak of the Second World War, Australia and New Zealand spent less than 1 per cent of their GNP on defence, while Britain was spending five times as much in relation to its GNP. India, previously Britain's vital source of manpower, began to oppose the use of its forces outside India, and after 1933 the British government had to pay for enlisted Indian forces deployed outside the subcontinent, making the imperial defence force much more expensive. With British society becoming more democratic after the 1918 Reform Act, and in the wake of the costly First World War, the British public demanded a reduction of the military burden and an increase in the provision of social security.[8]

These adverse circumstances did not immediately affect Britain's security interests *per se*, as the lessening of military threats after the First World War allowed Britain to adopt, in 1919, the Ten Year Rule, assuming that there would be no war on the Continent for ten years. Successive governments kept defence expenditures to a minimum. Britain monitored increasing Japanese and US naval capabilities which might eventually challenge Britain's naval power in the Far East. In 1919, Admiral Lord Jellicoe advocated the construction of a naval base in Singapore, but the cost was then regarded as too high to be worth proceeding with. However, in June 1921 Britain announced the construction of the Singapore base as a means of *deterring* a possible Japanese naval threat. The fleet could be sent to the base in an emergency. Aware of the United States' potential power, Britain had by then discounted the possibility of an exhausting and costly Anglo-American war. Instead it was decided that Britain's security lay in co-operation with the United States. South Africa and Canada also warned the British that maintaining the Anglo-Japanese alliance would harm its relations with the United States and the Empire. The Anglo-Japanese alliance lapsed when the five major naval powers, including Japan, the USA and Britain, concluded the Washington naval treaty in 1922.[9] Churchill recalled that 'it was with sorrow' that he supported the end of the alliance with Japan from which 'we derived both strength and advantage', but that, 'as we had to choose between Japanese and American friendship, I had no doubt what our course should be'.[10] A semblance of stability was also restored to Europe by the mid-1920s. The British government decided in 1928 to extend the Ten Year Rule on a rolling basis, and Britain postponed a number of defence projects, including the construction of the Singapore base. Germany was still bound by the Locarno and Versailles Treaties, and the post-1925 climate of British–French–German *rapprochement* would preclude Japan from allying with any major European power. Up to 1933 Britain persisted in the belief that Japan was unlikely to become an actual threat to Britain, while the Foreign Office was then more concerned about the growth of Chinese communism and its effect on the British presence in Shanghai.[11] During the early part of the inter-war period, Britain's grand strategy in peace sought to deter possible enemies by diplomacy and by the conclusion of treaties, and it was certainly most reluctant to start preparing militarily against potential or imagined threats.

Prioritising Britain's global security interests, 1931–41

After 1931, the international system was rapidly changing for the worse. Japan expanded into Manchuria, and created a Japanese puppet state, Manchukuo, withdrawing from the League of Nations in March 1933. Tokyo also walked out of the London Naval Conference on 15 January 1936. At the end of 1933, Germany, now led by Adolf Hitler, had also withdrawn from both the disarmament conference and the League, and these two powers were to be joined by Benito Mussolini's Italy in November 1937 to form the Anti-Comintern Pact, which, in September 1940, was developed into the Tripartite Axis Pact. The time had come for Britain to prioritise its security interests.

In 1934, the newly established Defence Requirements Committee chaired by Sir Maurice Hankey, the Cabinet Secretary, concluded that Germany, by virtue of its geographical proximity and its economic potential, was to be regarded as the 'greater adversary' than Japan. For the defence of the Eastern Empire, the Committee recommended the completion of the Singapore base (its construction had resumed in the previous year), but it hoped that Japan could, and should, be pacified by diplomatic means.[12] Accordingly, in the British rearmament programme, air defence was given the first priority, followed by the protection of seaborne trade and the defence of the Empire, while the expeditionary force came a poor third. It was also important to strike the right balance between defence and economy. Between 1934 and 1937, Neville Chamberlain, as Chancellor of the Exchequer, tried to ensure that rearmament did not affect the civilian sector of the economy or erode living standards. In fact Britain was spending only 3 per cent of its gross national product on defence in 1935, rising eventually to about 18 per cent by 1939, while German expended 8 per cent in 1935, increasing to 23 per cent prior to the outbreak of the Second World War. Japanese rearmament began in earnest in 1937, and by 1940 nearly half of the national budget was absorbed by military expenditures. This meant that in Japan by 1941 no oil or petrol was allocated for civilian use (cars had to be driven by coal-fired steam engines), while essential foodstuffs and goods (coal, sugar, matches, rice, salt and cotton) were severely rationed.[13]

Britain's priorities further sharpened during 1939–41. In June 1940, France surrendered (and Britain lost its only major ally in the European war), while Italy joined in the war on the side of the Axis powers and Hitler attacked the Soviet Union in June 1941. All of these factors impinged upon Britain's ability to prepare for the defence of Singapore, an issue which had been regarded as urgent by Australia and New Zealand for some time. The Royal Air Force had been built up to attack Germany, not to defend the Eastern Empire. To keep the new ally, the Soviet Union, in the war was another important consideration in the context of the war in Europe, as Britain initially underestimated Moscow's military capabilities in the wake of the Great Purges of 1937–38. Churchill was, despite his long-standing hatred of the Bolsheviks, determined to give 'all encouragement and any help we can spare to the Russians following [the] principle that Hitler is the foe we have to beat'. In September 1941, Britain's precious fighters were sent to Russia instead of to Malaya and Singapore.[14]

Perhaps the most important obstacle to Britain's Singapore strategy was the combination of France's defeat and Italy's entry on the side of the Axis. The former made it easier for Japan to expand into Indochina, which made war in the East more probable, while depriving Britain of France's naval strength to help contain Italy in the Mediterranean. Although the Singapore base had been finally completed in 1938, the British fleet was now concentrated in the Mediterranean. With an eye to the Italian threat, the Chiefs of Staff maintained in November 1939 that 'the sea route through the Mediterranean, the Suez Canal and the Red Sea to the East, was Britain's "primary" strategic interest in that theatre, followed by the Anglo-Iranian oilfields and the Indian Northwest Frontier'.[15] Italy's entry into the war threatened the second pillar of Britain's security. The defence of the Eastern Empire lagged even further behind. Britain's Singapore strategy was further undermined by the United States' unwillingness to defend Singapore and Malaya or help Britain in any way to protect the British Empire in the East.[16]

Between 1939 and 1941, Britain's security interests were prioritised, with the defence of Britain against a possible German invasion as the top priority. Within the British Empire, the defence of the Mediterranean/Middle East came first, followed by the defence of the Eastern Empire. Despite prodding from the Pacific dominions, Britain could not give a firm guarantee to defend Australia and New Zealand in the event of a major crisis in the East. Japan had in any case been engaged in war in China after 1937. Britain, sympathetic towards China's plight but primarily because of its trade and strategic interests in Shanghai and Hong Kong, sought to assist China. Of course, Britain had little interest in Manchuria, which, it believed, had never been an integral part of China. Many of the British leaders had little objection to Manchuria being governed by Japan which would have the obvious merit, in London's eyes, of diverting Japan from expanding into the Pacific.[17] In any case Britain was not involved in the detailed US–Japanese negotiations prior to the outbreak of the Pacific War. Britain regarded the question of resolving the Japanese–Chinese conflict as primarily the United States' affair.[18] In the wider context of Britain's global strategy, the Far Eastern conflict did not occupy a prominent position.

All these factors contributed to Britain's ambivalence about Japanese intentions in 1941, when Churchill and the Foreign Office believed that the Japanese would not attack Singapore before the Russians were defeated by Germany.[19] Indeed, it is difficult to ascertain when Japan finally decided to cross the Rubicon. Under the right-wing and expansionist Matsuoka Yosuke, the Japanese Foreign Minister, Japan concluded the tripartite pact in September 1940. Matsuoka also supported southward expansion, including an attack on Singapore. The Prime Minister, Prince Konoe, had not yet abandoned his hope for a negotiated settlement with the USA over Japan's sphere of influence in the Asia-Pacific region, and had decided to dismiss Matsuoka by dissolving his Cabinet in July 1941.[20] It was only in the autumn of 1941 that operational preparations for the Pacific War began in earnest. The Imperial Army completed its operational plans in September 1941, and embarked on full military training exercises in October 1941. The Japanese Navy had considerable difficulty in procuring the right kind of torpedoes, which were to play a significant role in the Pearl Harbor operation, and it was not until November 1941 that the Navy was able to work out how to operate these torpedoes successfully. If Churchill was banking on the assumption that Japan would not embark on a military adventure in South-East Asia, this does not appear unreasonable, since Japan's actual military decision to opt for war came very late after a great deal of muddled thinking and confusion in Tokyo.[21]

Britain's grand strategy with its allies – overcoming the differences

It was Churchill who shaped the largest decisions on the war, with an eye to the limitations of Britain's war resources. He was able to incorporate strategic variables at various levels into a broader context, each one related to others, and contributing, in varying degrees, to the formulation of Britain's grand strategy. Strong willed, bold and reckless at times, he stood by his principles, from which he never wavered throughout the war.[22]

1941 changed the strategic landscape for Britain. By the end of that year, the Soviet Union, China and the United States joined Britain as allies. The Second World War was now being fought by a global coalition. The most important ally for Britain was the United States, and Churchill's first task was to minimise the differences between the two powers, and to bring the United States firmly to the support of Britain's grand strategy. Throughout the rest of the Second World War, Anglo-American war strategy at the highest levels shaped the course of the global coalition warfare.

Churchill had once depicted US thought as 'large-scale mass production-style of thought' in that the United States was liable to become 'too short-sighted, too direct, blunt, too intent on military victory, too forgetful of the large objectives of war'.[23] Churchill assumed correctly that the United States, angered by Japan's surprise attack, would now want speedy action against Japan in the Pacific, which would result in the diversion of the United States' attention and resources from the war against Nazi Germany. Churchill met with the US President, Franklin Roosevelt, to discuss long-term war strategy at the Arcadia Conference in Washington between 22 December 1941 and 3 January 1942. Churchill and the British military leaders arrived in Washington well prepared and well briefed, and caught their US counterparts off guard. Britain had already been in the war for over two years, its troops were in place and it had a well-organised bureaucracy to support the nation's war efforts. The conference confirmed the Europe-first strategy and the defensive strategy in the Asian-Pacific theatre.[24]

The Europe-first strategy was by no means imposed by Britain on Washington. Franklin Roosevelt was an astute and enigmatic politician, who believed in the United States' international role and who sought to create a post-war new world order through collective security. The Roosevelt administration was naturally alarmed by the growing threats to the European democracies after the failure of the Munich settlement. The conclusion of the tripartite pact in September 1940 made the Roosevelt administration acutely aware of the possibility of the spread of the European war into Asia and the Pacific region. In 1941 the United States sought to ease the financial situation of Britain and later China by setting up the lend-lease programme. America's military planners produced numerous contingency plans, which envisaged priority being given to Europe, while undertaking defensive campaigns against Japan. In February 1942, Britain and the United States had agreed on the Europe-first strategy.[25]

Churchill and Roosevelt were therefore in complete agreement about the menace of Nazi Germany. The two men thought that the defeat of Germany would lead inevitably to the eventual defeat of Japan, but they doubted that the prior defeat of Japan would bring the Second World War to an end.[26] While the relationship between Churchill and Roosevelt became close and often warm for the rest of Roosevelt's life, it did not, of course, mean that the two countries agreed on the direction of the Allied war effort in all circumstances. Britain, because of its long experiences with international affairs, had formed clear opinions of, or prejudices against, certain countries. For instance, the Soviet Union was seen as tsarist Russia plus communist ideology accompanied by the same expansionist tendencies as the Russian Empire had exhibited. Britain was also inclined to see the solution to the international conflicts in terms of geopolitical settlements rather than collective security which was favoured by the United States, and retained somewhat paternalistic attitudes towards the newcomers in the international system like Japan. Churchill saw Japan with 'no background but the remote past', and as having been transformed into a modern industrial state under 'British and American guidance'. The United States and Britain were therefore, in Churchill's views, the 'god-parents' of the new Japan.[27]

Roosevelt regarded Churchill's political outlook as 'old-fashioned', ingrained with Victorian values, and he would certainly not have chosen him as his confidant had it not been for the Second World War.[28] The United States continued to be suspicious of Britain's imperialist interests, and its post-war intentions in Asia and elsewhere. More importantly, the United States was brought into the war because of Pearl Harbor, and had been traditionally reluctant to become involved in European conflicts. US public opinion and military leaders naturally regarded Japan as their direct enemy, and were anxious to defeat Japan as quickly as possible within the constraints of the Europe-first strategy.

Churchill's ideas about defeating Japan were rather different. In a memorandum, which was presented to the United States at the Arcadia Conference, the British Prime Minister maintained that Japan had long been 'overstrained by its wasteful war in China', and had been 'at their maximum strength on the day of the Pearl Harbor attack'. His prescription was therefore to keep the Japanese 'busy' in their occupied territories so as to deprive Japan of enjoying the fruits of its recent victories and prevent it from reducing its war effort. In order to do this, Britain agreed with the United States that the Allies should mount limited offensives against Japan wherever possible.[29] The outcome would be to exhaust the Japanese at a minimum cost to the Allies, thereby making the final campaign against Japan no more than a mopping-up operation, a typical exposition of Churchill's peripheral strategy. US rearmament only really started seriously in December 1941 – while Britain had already reached nearly 60 per cent of its maximum military output, the same figure for the United States was only 11 per cent.[30] It was therefore Britain that dominated the Arcadia Conference.

The Arcadia Conference also agreed to set up a global wartime decision-making apparatus with the establishment of the Anglo-US Combined Chief of Staff (CCS) to co-ordinate war efforts on a day-to-day basis and to harmonise Allied strategy.[31] For the Asia-Pacific region, it was agreed to establish ABDA (American, British, Dutch and Australian forces) as an umbrella command to cover the whole region threatened by Japanese aggression. Knowing that the Allies faced further disasters as the Japanese forces now had a free hand to occupy the theatre, neither Britain nor the United States wished to recommend one of its nationals to be appointed Commander of ABDA, a thankless task. In the end, Field Marshal Sir Archibald Wavell, the Commander-in-Chief in India, took the post, and, as anticipated, he witnessed a series of Allied defeats in the Dutch East Indies, Malaya, Singapore and Burma, thereby making ABDACOM utterly purposeless. It was dissolved at the beginning of March 1942, and was reconstituted as South-West Pacific Theatre Command under the flamboyant and controversial US general, Douglas MacArthur, in April 1942.[32]

The new ally, China, was of course delighted by the entry of the United States and Britain into the war. Chiang Kai-shek, the Chinese Nationalist leader, thought that China would now get the all-out support it needed from its allies. Chiang immediately proposed that an Allied military headquarters should be set up in Chungking.[33] Both Churchill and Roosevelt appreciated the importance of keeping China in the war, but, beyond that, Britain and the United States differed considerably over China. Britain had scant regard for China's military strength, had earlier been the target of China's anti-imperialism and was afraid of losing Hong Kong to China after the Pacific War. The United States had great expectations of China's potential role in the defeat of Japan. It believed – wrongly – that huge Chinese armies could be retrained and equipped for major offensive operations against the Japanese, while China's proximity to Japan would be useful for the Allied air campaign against the Japanese home islands. In addition, Roosevelt was hopeful that China would become a great power to counterbalance a post-war Japan. At the end of December 1941, the United States persuaded Britain and the Netherlands to agree to establish the China theatre, with Chiang Kai-shek as supreme commander. Subsequently, US General Joseph Stilwell was appointed as US Military Representative in China to improve the combat efficiency of the Chinese Army. Stilwell was also to be Chief of Staff to Chiang as well as commander of the US forces in China, Burma and India. The China factor, however, did not figure prominently in the initial Anglo-US combined war strategy.[34]

More important for Britain was to keep its Commonwealth and Empire together as part of the Anglo-US global coalition. While Britain's control of its Empire had begun to show some fragility during the inter-war period, it was able to obtain crucial military manpower from its Empire during the Second World War. Throughout the war, nearly half of Britain's armed forces were raised in India, West Africa and the Dominions. Imperial forces contributed significantly to Britain's defence of the Middle East and the Mediterranean between 1940 and 1942, since the bulk of the British national forces were engaged in home defence in case of a possible German invasion.[35] Before the outbreak of the Pacific War, Britain had had some difficulty in incorporating Australia into the defence of the Middle East, as Canberra insisted on an independent Australian army corps, separate from the British command.

Britain's relations with Australia were severely strained when the war in the Pacific broke out. Churchill was forced to agree to the withdrawal of the bulk of Australian formations from the Middle East to defend its own country. Singapore remained the key to the defence of the Eastern Empire. During the Arcadia Conference, the United States agreed to send troops and aeroplanes to Singapore in the event of the fall of the Philippines. By mid-January, British reinforcements of 9,000 troops, including guns and fighters, had arrived at Singapore. Soon after this reinforcement, it became clear to the British Prime Minister that the Singapore base was in fact far from being a fortress, since it was not completely fortified against an attack from the rear.[36] Three weeks later Singapore was lost to the Japanese.

Churchill's speech reporting the disaster to Parliament was not well received in Britain. The fall of Singapore was the result of a lucky combination of factors for Japan – the Japanese forces were suffering a shortage of ammunition and a loss of morale when Lieutenant-General Arthur Percival surrendered.[37] South-East Asia had been accorded the lowest priority in Britain's global strategy, and as such its defence was a calculated risk. The United States, when it entered into the war, was hardly in a position to field a large number of troops, and the Allies were able to muster only 12 divisions in the whole of the Pacific, out of which four had been lost in Singapore. The Japanese had fewer troops in Singapore, but enjoyed air and naval superiority over the Allies.[38]

After Singapore, the Pacific Dominions, feeling betrayed by the mother country, moved closer to the United States. Neither Britain nor the United States wanted these allies to interfere with their war strategy, much as they valued their contributions to the war effort. The establishment of the Pacific War Council in the spring of 1942 was a device for allowing smaller powers 'to let off steam', but it was never intended to be a decision-making body. In any case, the United States was determined that the Pacific War was to be its 'exclusive project', resenting interference from outsiders. In this climate, Australia was consistently prevented by Roosevelt and his military leaders from playing a major role in the Pacific War or in US post-war planning. While Canberra initially welcomed the creation of MacArthur's South-West Pacific Theatre Command in Australia, it soon became clear to the Australians that the new command would be MacArthur's power base.[39]

During the initial stages of the Pacific War, the division of labour was largely determined in accordance with British interests. Again both Roosevelt and Churchill were at one in their thinking about each other's role in the global war. By the end of March 1942, it was agreed that the United States would undertake the operational responsibilities for the Pacific area, while Britain would have similar authority in the 'middle area' from Singapore across the Indian Ocean, the Red Sea to the Mediterranean. Europe and the Atlantic would be a joint Anglo-US responsibility under the direction of the Combined Chiefs of Staff.[40] The problem was how the Europe-first strategy was to be actually carried out and, without knowing this, it was difficult to envisage when and how to defeat Japan.

The European factor

Churchill preferred a traditional peripheral strategy of gradually 'tightening the ring' round Germany, first clearing North Africa and opening the Mediterranean, but avoiding large-scale land operations against Germany until its morale and its ability to resist had collapsed.[41] His experience of the 1914–18 world war, and the recent Battle of Britain, seemed to justify his gradualist approach to fighting Germany. However, the US military leadership (General George Marshall and Admiral King in particular), confident of the United States' ability to build up its huge war machine quickly and anxious for a speedy defeat of Japan, wished to challenge the Germans on the Continent much earlier than Britain thought desirable. US military leaders felt that they were being manoeuvred by Britain into agreeing to a series of diversionary and sub-sidiary operations, while constantly delaying a cross-Channel operation. These differences became noticeable by the summer of 1942, and continued well into the end of 1943 when the Soviet Union, Britain and the United States finally agreed on the approximate timing of the cross-Channel operation – eventually known as Operation Overlord – to take place in the spring of 1944.

Clearly practical factors inhibited the launching of a cross-Channel operation in either 1942 or 1943, as King and Marshall wanted. There were not sufficient numbers of US troops in Britain, whereas Britain had more trained combat troops available there. Without Britain's agreement, therefore, the operation was impossible. On the other hand, both Roosevelt and Churchill were anxious to keep the Soviet Union in the war, and Stalin was constantly reminding the two allies of the urgent need for action in western Europe which would relieve the Soviet burden on the Eastern front. The worst-case scenario would be if a desperate Soviet Union made peace with Germany. It was therefore a political imperative for Britain and the United States to demonstrate to Moscow that they were also engaged directly in fighting the Germans. Domestic factors in Britain and the United States also played a part. Roosevelt feared that US public opinion, having witnessed Japan's earlier successive victories against the Allies in Asia and the Pacific, might demand a shift to a Pacific-first strategy. It was seen therefore as crucial for the Allies to mount a victorious campaign in Europe to concentrate the US public's attention on Europe and restore its confidence in the United States' military prowess. For Churchill a successful campaign in Europe would help to boost British morale, which had suffered as a result of a long series of reverses in Europe and Asia since the beginning of the Second World War. For these reasons, Britain and the United States agreed in July 1942 to launch an invasion of North Africa (Operation Torch) in the autumn.[42]

When the Anglo-US leaders met in Casablanca in North Africa in January 1943, both sides feared they might be forced to agree to operations with which one or other disagreed.[43] However, the same problem of resources and Roosevelt's anxiety to show Stalin that the Allies were ready to engage the Axis on land resulted in an Anglo-US agreement that the invasion of Sicily would be the next campaign (Operation Husky), whilst Anglo-US forces in Britain for the Second Front would be significantly increased.[44] As General Marshall rightly predicted, the campaigns in North Africa and the Mediterranean 'developed a life of their own, nourished by the strategy of the British Chiefs of Staff, while delaying [other] land operations against the Axis powers'.[45] The invasion of Sicily took place in July 1943, leading subsequently to the invasion of Italy. By mid-1943, more than half a million US troops and 110,000 British troops were in the Mediterranean. On 8 October, Italy surrendered.

The China factor

While the European war was making progress, Anglo-US disagreements over China, however, increased Allied divergences about the reconquest of Burma. The interplay between the China and the Burma theatres is worth noting, for Churchill and the British Chiefs of Staff were united in the view that the Burma theatre would be the worst place to fight the Japanese. The British Prime Minister preferred to bypass Burma and seize islands in the Dutch East Indies as a prelude to the recapture of Singapore.[46] In British eyes, Burma was last on the list of priorities in planning British campaigns against Japan. The United States, on the other hand, had no interest in helping Britain to reconquer its holdings in South-East Asia. In any case, the British had agreed that the United States would concentrate on the campaigns in the Pacific Ocean, leaving South-East Asia as Britain's responsibility. In the Anglo-US strategy for the war against Japan, Burma had become the forgotten theatre at the beginning of the Pacific War. But, as it turned out, it became a controversial strategic issue between the two powers by mid-1942.

With the fall of Burma, the Burma Road – a primitive 700-mile highway from Lashio in northern Burma into Kunming, China – was cut off. The loss of the Burma Road meant that China's only supply line was by an inefficient air transport system over the rugged Himalayan Mountains from India, which was vulnerable to interception by Japanese fighters. The amount of lend-lease provided to China between 1941 and 1944 was less than 1 per cent of the United States' assistance to all its allies. Only in 1945, with the expansion of the airlift and the reopening of the Burma Road, did the size of US aid to China increase significantly (ten times between 1944 and 1945).[47]

Chiang's Chief of Staff, Stilwell, was determined to retake northern Burma in order to ease China's supply problems. US military chiefs in Washington also favoured a major offensive into Burma in the spring or summer of 1943 for the same reason and as a means of deterring China from making peace with Japan. Chiang Kai-shek agreed to the Stilwell plan, but only on condition that Britain would mount amphibious operations to ensure the success of the plan.[48] Britain and the United States discussed this issue in August 1942, but British reactions were negative. Britain, at first suspicious of Sino-US intentions in Burma, declined to provide amphibious forces as they were fully engaged in the European war. The monsoon season precluded such an offensive taking place in the period the United States favoured. At Casablanca in January 1943, Britain agreed to the importance of proceeding to a major offensive campaign in Burma, codenamed Anakim, but, owing to the lack of landing craft, Britain claimed that Anakim could only take place towards the end of 1944. Admiral King described this decision as 'fantastic'.[49] For the British, the reconquest of Burma would be a hazardous undertaking and could not be contemplated until British–Indian forces on the border of Burma had been reinforced and re-equipped. Even if upper Burma was cleared of Japanese forces sufficiently to permit the building of a road into China, the defence of the road against the Japanese remaining in central Burma would be a major task. Only an amphibious operation (as Chiang was insisting) directed against Rangoon as the first stage of an offensive into central and upper Burma offered any realistic prospects of clearing Burma of Japanese forces.[50] Thus, the timing of a British amphibious operation became crucial to the maintenance of the Anglo-American–Chinese alliance.

Luck was, however, on Britain's side, as the whole idea of launching a large land operation in Burma was subject to delays. First, Chiang Kai-shek was reluctant to allow his armies to be retrained by US forces. Stilwell's aim was to create a combat-ready Chinese army of 30 divisions to reopen the Burma

Road, but he found that the Chinese Nationalist army were not prepared to obey Stilwell's orders, unless they came direct from Chiang Kai-shek. Stilwell, who spoke Chinese fluently, had a deep affection for the Chinese, but had no respect at all for Chiang, whom he (Stilwell) called, in his coded telegrams to the United States, 'peanut'.[51] The state of the Chinese armed forces was a matter of serious concern to the US general. Stilwell lamented that: 'the Chinese army lies immobile and rotting, sprawled all [over] China . . . officers getting rich, men dying of malnutrition, malaria . . . the sick simply turned loose . . . Stupidity, ignorance, apathy in the general staff . . . Personal loyalty to Chiang Kai-shek weighs more than ability and efficiency.'[52] Relations between Chiang and Stilwell became increasingly strained. After all, what Chiang really wanted from the USA was food and military supplies, and he expected Stilwell to work as his personal factotum to facilitate this process. Instead the United States wanted Chiang to use his army in support of the Allied war effort. China and the United States were never in full accord about each other's role in the war against Japan.

The second factor was a shift in Chiang's and US thinking about the use of air power from the Chinese mainland to bombard Japan. This was a pet scheme of General Claire Chennault, the leader of the American Volunteer Group, the 'Flying Tigers', in China. When the Pacific War broke out, Chennault's Tigers were incorporated into the regular US Army Air Force by Stilwell. The relations between the two men soured as they argued about their rival plans to defeat Japan. Chennault believed he could accomplish the destruction of Japan by air power alone. Chiang was attracted to this scheme since it would allow Chiang to conserve his army for action against the Chinese Communists after the war.[53]

These divisions in Chungking over future operations increased the rivalry between the US Army and Navy in Washington. General Marshall supported Stilwell's land operations in Burma, while Admirals Nimitz and King both supported the use of air power directly against Japan. Chennault, convincing and articulate, contrasted with 'Vinegar Joe' Stilwell's pessimism about Chiang. Roosevelt was equally attracted to a less costly air campaign against Japan. By May 1943, when Churchill and Roosevelt met in Washington, operation Anakim was no longer the United States' main concern, and it was quietly pigeonholed.[54] Thus, Chiang's stubbornness, Stilwell's difficulties in retraining the Chinese Army, and the eventual shift to air power all helped Britain to suspend an amphibious operation for the time being.

The bickering over China did not however leave the Anglo-US relationships without scars. If the United States was impressed with British negotiating skills and their approach to the war in Europe, this was not echoed by US officials and soldiers in South-East Asia, the heart of Britain's imperial sphere. They were at first surprised by British defeatism in the face of Japan's initial offensives. Thereafter Britain mounted an unimpressive and lengthy limited offensive in Arakan in Burma between the end of 1942 and early 1943, and the only bright spot was the guerrilla campaign of General Wingate's Chindits. Moreover the growing campaigns against imperialism in Burma and India further deepened US suspicions of the British imperial system there. There was also the US fear – as Churchill noted at the time of the Casablanca conference – that Britain might get out of the war altogether once Germany had been defeated, leaving the United States fighting alone against the Japanese.[55]

In order to demonstrate to the United States Britain's seriousness of purposes in the war against Japan and to restore Britain's image in South-East Asia, London proposed the creation of the South-East Asia Command (SEAC) under Vice-Admiral Lord Louis Mountbatten in the summer of 1943. Moutbatten's tasks were to clear the Japanese from Burma, Malaya and the rest of South-East Asia, and to reopen the land communications with China across the north of Burma.[56]

Mountbatten, a cousin of King George VI, and the youngest vice-admiral in the Royal Navy, was 42 years old when he was appointed Supreme Allied Commander, South-East Asia, in August 1943, with General Stilwell as his deputy. Many US officers sent to SEAC felt morally uncomfortable in that they might be regarded by Asians as assisting the British to restore their empire in South-East Asia, and they branded the SEAC as 'Save England's Asiatic Colonies'.[57] The SEAC remained relatively low on the Allied list of priorities. Mountbatten was often deprived of resources and men to achieve his military objectives. It was perhaps one of the most frustrating and politically divided theatres of all the theatres in which the United States and Britain jointly fought.

In late November 1943, the Anglo-US leaders invited Chiang Kai-shek to Cairo to discuss Asian strategy. Mountbatten was also invited to attend as the new SEAC Commander. Chiang was once again assured of the Allied determination to reopen the Burma Road, and Mountbatten offered, to the surprise of Churchill, an amphibious operation in the Bay of Bengal, Buccaneer, as part of this operation. Unlike the previous Washington conference, Roosevelt was clearly intent on pushing the British to embark on an amphibious operation in South-East Asia. Nor was Roosevelt prepared to agree to any further Allied operations in the Mediterranean as Churchill contemplated. The US President wanted to concentrate Allied efforts solely on Overlord. Anglo-US leaders then moved to Tehran in December to meet Stalin, who strenuously objected to any further postponement of Overlord. To the relief of the Anglo-US leaders, the Soviet leaders agreed to enter the war against Japan after Germany was defeated.[58]

Stalin's pledge to join the Allied effort in the Far East had deep implications for Anglo-US war strategy. Roosevelt began to realise that China was not likely to be of much use militarily in the war against Japan, and he was delighted by Stalin's proposed assistance. Having said this, Roosevelt was still keen to keep China in the war, and was possibly thinking about the creation of a post-war US–Soviet–Chinese 'entente' in Asia, to the exclusion of Britain. Churchill took Stalin's assurance of participating in the Pacific War to mean that Chinese military power was no longer necessary for the defeat of Japan, thereby reducing the urgency of launching Buccaneer.[59] By then differences between Roosevelt and Churchill were increasing. Roosevelt's mind was becoming focused on Asia after the defeat of Japan, whilst Churchill's attention was constantly on post-war Europe where he was anxious to contain possible Soviet expansionism in central and southern Europe. The British continued to be preoccupied with the idea of weakening the Germans wherever possible outside western Europe at the cost of delaying the timing of Overlord – much to the annoyance of the US leaders.

The problem was again the question of securing sufficient landing craft. The two Western allies now expected both Overlord and Anvil (a landing in southern France) to take place in mid-1944. Meanwhile, Churchill wanted to continue to weaken the Germans, preferably in the Balkans and eastern Mediterranean, by persuading Turkey to enter the war on the Allied side. All the above operations required landing craft. Roosevelt preferred the use of landing craft for Buccaneer over Churchill's proposed eastern Mediterranean campaign. Churchill however reminded the President that Anvil, too, required landing craft, and insisted that it made sense to keep the landing craft in the eastern Mediterranean until nearer the time when Overlord/Anvil operations were to be carried out.[60] Once again, the Europe-first strategy diverted resources from the war against Japan, and Churchill persuaded Roosevelt during the Anglo-US conference back in Cairo in December 1943 to cancel Buccaneer.

The defeat of Japan

Stalin's pledge, and the cancellation of Buccaneer, in the end wiped out any remaining opportunity for China to play an active part in the war against Japan. Chennault's dream of using China as an Allied air base was also crushed by Japan's Ichigo operation in April 1944, designed to capture Chennault's airfields. This was the largest Japanese campaign in the war in China, involving 620,000 troops. By then, the United States had decided that, given the progress being made in its campaign in the Pacific, the main effort to defeat Japan would be made through the Central and South-West Pacific. In Burma, Mountbatten's Allied offensive called Capital across the Chindwin River from India into northern Burma was delayed until October 1944 by Japan's disastrous campaigns in Imphal and Kohima. In April 1945 Britain's long-awaited amphibious expedition from the Bay of Bengal, supported by General Slim's Fourteenth Army advancing from Meiktila and Mandalay, led to the recapture of Rangoon, although the Japanese had already abandoned the capital.

As the European war was gradually coming to an end, Churchill and the British leaders were anxious to show solidarity with the United States in the Pacific War and to share the 'final victory' in the war against Japan. There were 160,000 British prisoners of war and civilian internees were in the hands of the Japanese. Britain's aim was actually to participate in the final battle against Japan, and not just to liberate British dependent territories in South-East Asia occupied by Japan. During the second Quebec Conference in September 1944, Churchill suggested that the British main fleet should be placed under the US Supreme Command in the Pacific, while the Royal Air Force should take part in the bombardment of Japan. Roosevelt enthusiastically accepted Churchill's offer, although some of the US military chiefs did so somewhat grudgingly, as they felt that they could now defeat Japan single-handedly.[61] The British Pacific fleet was subordinated to Admiral Nimitz's Central Pacific Command under the name of Task Force 57, and included four aircraft carriers, two battleships, five cruisers and 14 destroyers.[62] The British carrier force was involved in the Battle of Okinawa after April 1945, and witnessed the end of the famous Japanese battleship *Yamato*. The British fleet then bombarded the island of Miyako on 4 May until its ammunition ran out. In a message to Churchill on 25 May 1945, US Admiral Reymond Spruance stated his appreciation of British co-operation, stating that 'Task Force 57 has mirrored the great traditions of the Royal Navy to the American Task Forces'.[63]

The final campaign against Japan's mainland was discussed during the Yalta conference in February 1945. The United States estimated that it would take 18 months between the surrender of Germany and the defeat of Japan. Stalin had already made clear his specific claims in China and in Japan in return for the Soviet Union's participation in the Pacific War. At Yalta, Washington was glad that Stalin confirmed the Soviet Union's intention of entering the war against Japan two to three months after the surrender of Germany. Britain was not involved in the Stalin–Roosevelt discussions about the peace settlement with Japan, but Churchill approved the US–Soviet accords and signed them. Britain also appreciated the Soviet Union's participation in the Pacific War, since it might help divert Japanese troops to Manchuria from Malaya which Britain was planning to recapture.[64]

For the final campaign against the mainland of Japan, Britain was prepared to contribute some three to five divisions and 20 air squadrons to be built up jointly with the Dominions, with the first operation against Kyushu beginning in November 1945, followed by the invasion of Honshu in 1946. However, events were overtaken by the successful testing of the atomic bomb during the Potsdam conference. Having encountered the fanaticism of the Japanese forces in Okinawa and the increasing use of suicide bombers (kamikaze), Churchill hoped that the atomic bomb would offer the 'vision . . . of the end of the whole war in one or two violent shocks', and that is what happened. Churchill's immediate thought was that 'the Japanese people, whose courage I had always admired, might find in the apparition of this almost supernatural weapon an excuse which would save their honour and release them from their obligation of being killed to the last fighting man'.[65] Japan surrendered unconditionally on 15 August 1945.

At the end of August, British troops were still on their way to Singapore by sea, and it was not until 12 September 1945 that Lord Mountbatten received the formal and unconditional surrender of all Japanese forces within his South-East Asia Command. Britain's war against Japan was now over.

Conclusions

Britain's grand strategy for the war against Japan had been part of Britain's overall grand strategy for the Second World War. Britain used only a meagre portion of its resources and manpower in South-East Asia. All Britain wished to do in the theatre was to fight when opportunities arose, but not at the cost of the British war effort in Europe, North Africa and the Mediterranean. When the end of the war approached, Britain wanted to liberate its dependent territories occupied by Japan, and not 'have them handed back to [them] at the peace table'.[66]

The fact that Britain's interest in the Far East was so clear cut, and did not go beyond certain countries in South-East Asia, had been an ongoing process. After the British Empire reached its maximum extent in the wake of the First World War, Britain wanted to avoid extending its responsibilities any further, as they were becoming expensive and unaffordable. Moreover, by the early 1920s Britain's naval supremacy *vis-à-vis* the United States had ended, and it was during this time that Britain decided to co-operate with the United States in the Asia-Pacific region. From this it followed that Britain's differences with the United States over Manchuria or China or Japan would not become sources of severe Anglo-US quarrels. Given the United States' strength, it was reasonable to assume that it would pull its weight and prevent any single dominant power from upsetting the balance of power in the Asian-Pacific region. After all, the country was facing Asia across the Pacific Ocean. Britain's rather detached attitude towards the Japanese issue was demonstrated by its lack of participation in the US–Japanese negotiations prior to the war, or in the US–Soviet wartime discussions about post-war Asia. Britain had no qualms about the fact that the United States wanted to be the dominating partner in the prosecution of the Pacific War.

Grand strategy is a strategy for the survival of a state in war and peace. Britain was intent on maximising its resources for the preparations and execution of war. War was for Britain, to quote Clausewitz's famous axiom, 'the continuation of politics by other means'. If Britain was compelled to go to war, it did not mean that Britain would fight at all costs with no regard to the post-war world. It is in this context that Britain found Japan a puzzle.

The threat from Germany was more strongly felt by the mid-1930s, as Britain knew that Germany was capable of challenging the security of the European Continent if it so intended. The First World War was still vivid in the minds of the decision makers. Moreover the growth of air power placed the homeland of Britain under the threat of direct attack. Japan, on the other hand, a rapidly modernising and young country, was far away from Britain. While Japan's expansionist tendencies worried the British leadership, they never believed, as it turned out correctly, that Japan possessed the capability, in addition to its fighting in China, to win a war against the combined strengths of Britain and the United States. In theory, it was a puzzle as to why, without the necessary capability, Japan should decide to embark on such a struggle. Japan was for Britain an asymmetrical enemy, which was prepared to use suicide bombers, to choose death as an alternative to life and to waste the wealth, manpower and infrastructure it had built up for the sake of fighting an unwinnable war.[67] All of these were beyond Britain's comprehension, as its norm was to conserve as much as it could, while making every effort to maximise its resources.

Accordingly, although the Second World War further exhausted Britain economically, the sense of continuity remained strong. The war had left Britain's institutions and its parliamentary system largely intact, while the rest of Europe was in ruin and in despair. The first post-war government led by Clement Attlee was committed to the maintenance of full employment and to far-reaching reforms in education, health and welfare. Although Britain was now further behind the United States in the economic league table, it was still the second largest global power in the West.

Britain's keen interest in recovering its South-East Asian territories does not mean that it intended to re-establish the Empire as it had stood in 1941. After the war, Asia continued to receive a low priority in the formulation of Britain's global policy (perhaps except for a few years during the Korean War). The Middle East and Europe were paramount in Britain's security thinking. Attlee's Labour government was aware that Britain's traditional imperialism was becoming increasingly outdated. The writing was already on the wall in the case of India and Burma prior to, and during, the Pacific War. These countries were granted independence, together with Ceylon, by 1948. However, it was remarkable that Britain kept its informal Empire in the form of the Commonwealth (India, Pakistan, Singapore and Malaya included) in South-East Asia as late as the early 1970s, and it was Britain's decision in 1967 to withdraw its military forces from its bases in Singapore and Malaysia which marked the end of Britain's permanent military commitment east of Suez (South-East Asia). This decision was subject to strong protests by the United States, Australia, New Zealand and Singapore, all of which begged Britain not to abandon South-East Asia. In the mid-1960s, the United States, whose commitment to Vietnam was growing, told Britain frequently how much the United States valued Britain's policing role in South-East Asia, as Washington would be unable to take Britain's place there if it left.[68] These facts are certainly at odds with the familiar assertion that Britain lost its Empire as a result of the Pacific War. The failure or success of Britain's grand strategy in the Asia-Pacific war must therefore be seen in the global context and in the long-range historical context.

2

TOJO HIDEKI AS A WAR LEADER

Ryoichi Tobe

'War Lords Anonymous'

In his book *The War Lords*, A. J. P. Taylor deals with Japanese war leaders differently from those of other powers. Each chapter on those powers is given the title of their leader's name, for example, 'Mussolini', 'Hitler', 'Churchill', 'Stalin' or 'Roosevelt'. But the chapter on Japan is entitled 'War Lords Anonymous'. As Taylor observes, 'there was no Japanese war lord – no single figure who led Japan into war, who directed the war, who made the decisions, and so on'.[1]

On the cover of the book, however, we can see the portrait of Tojo Hideki, which is depicted together with the other five war leaders. If there was anyone who led Japan in the Second World War, it was Tojo, even if he might not deserve the title of war leader.

In fact, Tojo was the War Minister and the Home Minister in addition to being the Prime Minister at the outbreak of the Pacific War, and afterwards he held the additional posts of Munitions Minister and Chief of the Army General Staff. He had so many important posts and so wide a jurisdiction that it would be strange not to place him as a war leader on a par with the other five leaders. Moreover, Tojo was a full general on active duty. He was a professional soldier, but the other five were not. Therefore Tojo was more familiar than they with the military aspects of directing the war. In this respect, he is well qualified as a war leader.

It is true that Tojo was not head of state. But the Emperor as head of state hardly exerted predominant influence over particular decisions about directing the war. His expected role was to ratify the decisions made based on consensus among all concerned.[2] It was Tojo who made efforts to achieve consensus and who was responsible for making decisions on it.

Nevertheless Taylor does not put Tojo in the same category with the other five. He does not think that Tojo could well claim the name of war leader. Such an image of Tojo as a defective leader in wartime reflects the realities of war leadership in Japan. We will inquire into the characteristics of Japan's war leadership focused on Tojo, as far as Japan's war against Britain is concerned.

Institutional constraints

Viewed in terms of institutions, the reason why Tojo is not regarded as a proper war leader is rather simple. First, he could not direct military strategies or operations until he became Chief of the Army General Staff. It was more serious that Tojo hardly played a part in naval matters. The other five war leaders often played leading roles in devising and directing military strategies and operations. They participated in matters concerning all services: army, navy and air force. Thus, Tojo could not stand comparison with them.

It goes without saying that the Independence of Supreme Command was institutionally most responsible for constraining Tojo as a war leader. Putting it briefly, the institution meant that only the Army and the Navy General Staffs dealt with military strategies and operations. In November 1937 after the outbreak of the China War, the Imperial General Headquarters (IGHQ) was re-established in order to arrange the military aspects of directing war. It consisted of the senior officers from two General Staffs and the War and the Navy Ministries. At the same time, the Liaison Conference between the IGHQ and the government was established, which was expected to integrate the political aspects and military ones in the conduct of war. But it could not fulfil the expected role. All that the Liaison Conference did was literally 'liaison' between the two institutions. The civilian leaders were not able to take part in military strategic matters.

It is rather significant that Tojo held the post of War Minister concurrently with the premiership. He was the only Prime Minister who had simultaneously held the post of War Minister since 1885 when the cabinet system was introduced into Japan. Usually a military officer who was appointed Prime Minister would retire from the service. The War Minister had to be an active officer. But when Emperor Showa appointed Tojo Prime Minister in October 1941, he allowed Tojo to remain on the active list so that Tojo as War Minister could restrain the Army's frustration in the event that Japan did not go to war with the United States and Britain. So Tojo as the War Minister was able to participate in the IGHQ, which was otherwise impossible for the Prime Minister.

Tojo, a member of the IGHQ, was informed beforehand of the Navy's plan for the Pearl Harbor attack.[3] He could not have known the plan in advance if he had not been the War Minister. Since he was a member of the IGHQ, he could be informed of the strategic plans of the Army and the Navy, and receive reports on the military situation. In this sense, he was able to participate in the consultation of military strategies and operations. But it is one thing to participate in the consultation, and another to play a leading role in it. The War Minister could attend the meetings of the IGHQ, but could not take the initiative in or decide on military strategies and operations. Eventually Tojo resorted to seizing the post of Chief of the Army General Staff, because he could not direct the war just as War Minister and Prime Minister.

The Independence of Supreme Command was institutionalised in the late 1870s. With this institution, Japan fought two wars against China and Russia in the Meiji era. It did not seriously hinder the conduct of the two wars. During the Sino-Japanese War, Prime Minister Ito Hirobumi attended the meetings of the IGHQ, and got involved in military operations, notwithstanding he was a civilian. During the Russo-Japanese War, Prime Minister Katsura Taro, a retired general, theoretically could not take part in military strategies and operations, but he attended the important meetings of the IGHQ together with the elder statesmen, and played a leading role in conduct of the war.[4]

In other words, Ito and Katsura disregarded the Independence of Supreme Command when the institution would have hindered the effectiveness of their war leadership. They accomplished excellent war leadership by disregarding the institution when it was necessary to do so. Why, then, could Tojo not disregard it?

We can indicate the following points. The Meiji leaders who established the Independence of Supreme Command knew well the original purposes of the institution, so they did not hesitate to disregard it whenever they faced the situations irrelevant to it. In contrast with them, the military leaders in the Showa era were brought up with the institution. Their autonomy and prerogatives were guaranteed by it. Therefore they, including Tojo, were constrained by it.

It was very difficult for Tojo to disregard the Independence of Supreme Command. The military establishment would oppose him if he tried to disregard it. He himself would find it difficult psychologically to disregard it. Thus Tojo did not try to overcome the constraints until the last stage of his administration. He endeavoured to be a war leader within the limits of the institutional constraints.

Tojo's leadership style

We can find the characteristics of Tojo's leadership style in certain episodes. In October 1941 when he set about organising a new Cabinet, he excluded his military staff from his Cabinet organisation headquarters.[5]

That episode is presented as indicating Tojo's sincerity. He set store by making distinctions between the Prime Minister (civilian) and the War Minister (soldier), between military administration and operation, and between public life and private one, etc. The episode is also interesting as indicating his basic thinking on the relationship between political affairs and military ones. He not only refused to allow political matters (or partisan considerations) to intrude into military affairs, but also tried to prevent military considerations from affecting political matters. Tojo endeavoured to be a war leader on the basis of separating political matters from military affairs.

However, it was hardly possible to direct the war on the basis of the separation. How, then, did Tojo try to integrate political affairs and military ones? He adhered to formality in a bureaucratic sense. Putting it simply, he made great efforts to perform two parts separately, the part of Prime Minister and the part of War Minister. Tojo even executed his business partly in the office of the Prime Minister, partly in that of the War Minister.[6] He tried in earnest to play the role of Prime Minister at certain times, and that of War Minister at other times. He did not integrate political and military aspects of war leadership by concentrating and fusing the powers of the two posts into one.

There is another interesting episode symbolising Tojo's leadership style. In spring 1942 Tojo gave the following instruction to Sato Kenryo, the newly appointed Chief of the Military Affairs Bureau in the War Ministry:

> Conflicts between the Army and the Navy should be solved by co-ordination between you and your counterpart in the Navy Ministry. If those conflicts are not solved at the Bureau Chiefs' level, they will be brought to the War Minister and the Navy Minister for settlement. And if they are not solved at the ministerial level, ultimately the Prime Minister will have to mediate between the two services. But such mediation will be impossible when the Prime Minister is also the War Minister as I am. In such a case, the two services will be split and antagonistic towards each other.[7]

The other five leaders mentioned above would not have given Sato such an instruction. They would have solved any troublesome conflicts that might split two services themselves. They would not have left the solution to their subordinates. They would have enforced the solution and suppressed the split. Tojo, however, did not think in this way. He tried to leave troublesome conflicts to his subordinates' co-ordination. He would avoid as far as possible friction at ministerial level.

Tojo's stand to avoid confrontation was not only taken toward the Navy, but also toward the Army General Staff. He hesitated to complicate conflicts by disputing directly with the General Staff even when he had different opinions from it. He would not challenge the Independence of Supreme Command. He attached importance to the distinction between military administration and operation. All he would do was take indirect measures in a roundabout way to restrain the General Staff when it went too far. A good instance is the step he took to withdraw from Guadalcanal against the General Staff's wish to continue the campaign on the island.

At that time the War Ministry did not object to continuing the campaign, but tried to induce the General Staff to stop it and withdraw, by limiting the tonnage of transport ships required by the General Staff. Tojo and his government including the War Ministry did not directly challenge the General

Staff, or meddle in military strategic and operational matters. They tried to restrain and check the General Staff with administrative measures. Tojo avoided direct confrontation with it. He took a roundabout way to restrain it. But the officers in the General Staff regarded it as an increased interference by the War Ministry and were opposed to it.[8]

Nishiura Susumu, who was Secretary to the War Minister and then appointed Chief of the Military Affairs Section in the War Ministry in April 1942, commented on the relationship between the War Ministry and the General Staff. According to him, there were three kinds of cases where the War Ministry might differ from the General Staff on operational plans. First, the War Ministry might think that the operational plan was impossible to carry out because of limited resources such as budget, materials and personnel. Second, the operational plan might not be in accord with the Cabinet's policy. Third, the War Ministry's officers might not agree with the operational plan from the military professional viewpoint. In the first case, the War Ministry could not object to the operation if the General Staff understood the limited resources and insisted on carrying it out in spite of them. In the third case, the officer in the War Ministry might privately convey his doubts about the operation, but traditionally did not enforce his opinion. Only in the second case, then, was the War Ministry allowed to object to the operation and to force the General Staff to stop it.[9] Tojo as the War Minister seemed to think in the same way as Nishiura. The controversy over Guadalcanal was the first kind of case.

Japan was beaten at Midway and Guadalcanal, and suffered defeat after defeat after those battles. Both the Army and the Navy General Staffs planned new operations to restore falling fortunes, and demanded the requisition of the tremendous tonnage of transport ships required for the operations. When a large number of transport ships were requisitioned, there remained too few ships to transport important resources from the occupied regions in South-East Asia to Japan, and then building up national power including production of munitions faced difficulties. In this way the gap between directing the war and executing the military operations was widened increasingly with Japan's falling fortunes. Imoto Kumao, the Secretary to the War Minister appointed in October 1943, heard Tojo complaining many times of the Independence of Supreme Command: 'It is impossible to direct the war under the Independence of Supreme Command.'[10] His complaint should have fallen under the second case of Nishiura's categorisation, but Tojo did not force the General Staff to accept his objection.

When his frustration and dissatisfaction with the General Staff reached a peak in February 1944, Tojo resorted to seizing the post of Chief of the Army General Staff. But it is doubtful that this measure worked well. As usual Tojo executed his business partly in the office of the Prime Minister, partly in the office of the War Minister and partly in the office of the Chief of the General Staff. Some officers appreciated Tojo's new status because it made the Army's office work go on smoothly.[11] Certainly that might be true. But the political aspects and military ones of conducting the war were not integrated. Tojo could not play a proper role of war leader by this measure. In fact, he did not concentrate powers and authorities in himself, but just performed the three roles separately. He did not try to integrate political affairs and military ones, but tried to avoid the conflicts between them and to keep the office work going.

Strategy for defeating Britain

How did Tojo direct the war against Britain? First, we have to understand that defeating Britain was a key to ending the war in Japan's grand strategy.

The grand strategy was formulated just before the war. This stated that: 'Japan will first take a position of strategic advantage by destroying the enemy's bases in East Asia and the South-West Pacific, aim at lasting self-sufficiency by securing important resource areas and major communication lines, and establish a long-term invincible posture.' But Japan could not end the war in this way, let alone win it. Japan, therefore, incorporated into its grand strategy that it would make the United States lose the will to continue the war by defeating China and Britain in co-operation with Germany and Italy.[12]

Tojo seemed to agree with the strategy. He stated just before the war:

> Japan can hardly end the war in a short time because it has not the means to seal the fate of the enemy. The probability that the war will be long may be 80 per cent. But, the war could be short only in the following cases: (a) Japan destroys the US Navy main fleet (this may be successful if Japan captures the Philippines and the US fleet comes to regain it); (b) the United States loses the will to fight because Germany declares war on it or lands in England; (c) the attitude of the United States toward the war is changed by Axis commerce raiding warfare against Britain which drives it into a critical situation.[13]

Both cases (b) and (c) above aimed at making the United States lose the will to fight by causing Britain's collapse. Japan had to depend upon German actions in order to defeat Britain and to make the United States discontinue the war.

The US Navy's main fleet, except its aircraft carriers, was destroyed in Pearl Harbor at the opening of the war, though not in the same way as Tojo expected (he was informed of the surprise attack just a week before). But Japan could not have prospects of ending the war in a short time. Undoubtedly, the objectives in the first stage of Japan's grand strategy were fulfilled. Japan destroyed the enemy's bases, took the position of strategic advantage and secured important resource areas and major communication lines. At this time the Army tried to establish a long-term invincible position, and to adopt a strategic defence posture according to the plan made before the war. But the Navy began to insist on continuing the offensive in the Pacific, with the gathered momentum of the victories in the first stage. The Navy appeared to aim at ending the war by a showdown with the US Navy. In contrast to the Navy, the Army emphasised again the existing plan aiming at ending the war by causing China's and Britain's defeat.

In this controversy between the two services, Tojo suggested a reconsideration of the existing grand strategy. At the Liaison Conference in February 1942, he stated, 'It is now necessary to study hard how to conduct the war in the future from the viewpoints not only of the military, but also of the nation as a whole.'[14] At his suggestion, the IGHQ began to consider a new grand strategy.

Why did Tojo suggest such a reconsideration? There is no doubt that he agreed with the Army's strategy that Japan should give priority to the establishment of an invincible position and to the defeat of Britain. But Tojo calculated that Japan's victories in the first stage were of greater value than he had expected.[15] He probably sought new plans to defeat Britain to take full advantage of these early victories. By presenting new plans, he tried to convince the Navy that it should not deviate from the basic lines of the existing grand strategy. Tojo expected that the capture of Burma would have some effect on causing China's collapse, and hoped that Japan would then be able to advance into West Asia.[16]

However, the new grand strategy formulated in March 1942 was an ambiguous agreement that put the Army's and the Navy's claims side by side. It stated: 'Japan should build on the fruits of victories, and establish a political and military long-term invincible position, as well as seize an opportunity to take positive measures to cause Britain's collapse and make the United States lose the will to fight.'[17] This could be interpreted in two ways: (i) as establishing a long-term invincible strategic defence, and (ii) as seizing an opportunity to fight a decisive battle with the United States. In fact, both the Army and the Navy interpreted it in accordance with their own position. Tojo attempted to obtain reconfirmation by the two services of the basic line of the grand strategy, but did not succeed. And when the two services made an ambiguous agreement, he did not take the initiative to clarify it.

In June 1942, it seemed possible to make the strategies of the Army and the Navy converge. The Army guessed that the Navy would return to the strategy that gave priority to defeating Britain in the Indian Ocean, because the Navy had abandoned the idea of a showdown with the United States in the Pacific after the defeat at Midway. Besides, at that time, on the North African front, the Axis forces led by Erwin Rommel had just captured Tobruk, the strategic point in Libya, and invaded Egypt. Earlier the Army had studied a grandiose plan that the Axis forces traversing Egypt and advancing eastwards via Suez would join up in West Asia with the Japanese forces advancing westwards via India. This was called the 'Operation of Going Through India to West Asia'. The Army had expected much of this operation as a measure to cause Britain's defeat, but afterwards postponed it because it was predicted that the advance of the Axis forces into West Asia would now not begin until 1943 or after. Now, in mid-1942, the grandiose plan postponed earlier seemed to be a reality.

The Army took great interest in the development of the campaigns by Axis forces in North Africa. Tojo took a special interest.[18] Before the war began he had argued for the possibility of acting in concert with the Axis forces advancing into the Caucasus or the Near East.[19]

In mid-1942, the Army General Staff studied the Indian Ocean operation (the operation to capture Ceylon), the Chungking operation and the operation to advance into India. Of the three, Tojo was most positive about the Indian Ocean operation. It was unusual for him to be so positive, in view of the cautiousness with which he viewed the Chungking operation. The Indian Ocean operation was deemed to be a combined one by the Army and the Navy, part of the advance into West Asia. As Tanaka Shin'ichi, the Chief of the Operation Division in the Army General Staff, recalled after the war, 'Tojo seemed to be satisfied with the appearance that the Navy accepted again the strategy of defeating Britain and that the strategies of the Army and the Navy had converged.'[20] That episode shows clearly how much importance Tojo attached to co-operation, or avoidance of conflict, between the Army and the Navy.

But the Indian Ocean operation was not realised after all. In August the Navy returned to war with the United States and fought US forces advancing into the Solomon Islands. Before long the Army also got involved in the battles at Guadalcanal. At first Tojo was positive about capturing Guadalcanal. However, as the Japanese forces were defeated in battle after battle on the island, and their supplies fell into an extremely critical state, Tojo began to lean towards withdrawing. As mentioned above, his step to limit the tonnage of transport ships produced conflicts with the General Staff.

When the Japanese forces were losing the Guadalcanal campaign, the possibility of defeating Britain almost disappeared. The situation on the North African front was reversed after the battle at El Alamein. Japan lost the power to venture large-scale operations in the Indian Ocean or into the Indian subcontinent. At the Liaison Conference in late February 1943, Tojo questioned the appropriateness of the strategy of seeking to end the war by causing Britain's collapse.[21] His question hit the mark, but he did not present alternatives to the existing strategy.

The questions and doubts that Tojo posed at the Liaison Conferences and at the meetings of the IGHQ were pertinent in most cases. Good examples are his suggestion of reconsidering the grand strategy after the early victories, and his question about the strategic plan of ending the war by defeating Britain. But it is significant that he did not offer new strategic visions or guiding principles for directing the war. Even when the reconsideration set about by his suggestion went a different way from what he had intended, he did not try to correct its course.

India and Burma

In the course of 1943, Japan abandoned the strategy of defeating Britain. Japan's war plan was compelled to change. This may be an exaggeration, but it lost the prospect of victory, because the Axis forces were unable to effect Britain's collapse, and because the strategic environments worsened with the progress of the full-scale US counter-offensive in the Pacific. From the outset, Japan's grand strategy had aimed at securing a long-term invincible position. It could not expect to defeat the United States by itself. This implied that Japan could win the war only when, in co-operation with the Axis powers, it defeated Britain and caused the United States to lose the will to fight. When it proved impossible to cause Britain's collapse, it became extremely difficult to win the war. Japan, therefore, increasingly laid emphasis on avoiding defeat rather than on winning the war.

The Imperial Conference in late September 1943 decided a new grand strategy in the context of the worsening military situation in the South Pacific and Italy's breaking away from the Axis. The new strategy defined an 'Absolute Defence Sphere' as the area to secure to the last. The outer line of the area extended from the Kuriles, through the Bonins, to the mandated islands in the South Sea, the western region of New Guinea, the Sundas and Burma.[22] The significance of avoiding defeat was stressed more than ever. The new strategy included a plan to gain the confidence of the native residents in the occupied areas in South-East Asia, in order to strengthen their collaboration with Japan. It was Tojo who added this plan to the draft of the new strategy. He thought it necessary to counter the enemy's political propaganda, which was expected to increase with its counter-offensive.[23]

Tojo assumed a positive attitude toward 'political strategy' in the sense of gaining the confidence of the peoples in South-East Asia from the outset. He declared that Japan would give independence to Burma if the Burmese supported Japan's war, at the beginning of 1942 when it started the operation to capture Burma. This was a part of the 'political strategy' to isolate Britain from India and Burma by promoting Burmese independence and instigating the Indian independence movement. In April 1942, immediately after Stafford Cripps visited India to demand its co-operation with the British war effort, Tojo appealed in the Imperial Diet that an India ruled by the Indians would be realised soon after a Burma ruled by the Burmese. He warned the General Staff not to bomb India during this politically delicate situation.[24]

The emphasis of 'political strategy' was being transferred from defeating Britain to counteracting the enemy's counter-offensives in 1943. Tojo visited the Philippines in May, and Thailand, Singapore and Indonesia in June and July 1943. He was the first Japanese Prime Minister ever to visit South-East Asia. Japan recognised the independence of Burma and formed an alliance with it. Although Tojo did not visit Burma, he had five conversations with Ba Maw, a leader of Burma, more than with any other leader in South-East Asia.[25] Tojo positively supported Subhas Chandra Bose, the militant leader of the India independence movement, who came to Japan by submarine from Germany. In October, Japan recognised the provisional government of Free India, and decided that the Andaman and the Nicobar Islands would be granted to the Indian government on the day after the Greater East Asia Conference opened in November 1943. Tojo, having accepted Bose's request for the territories, immediately put the proposal before the Liaison Conference, which approved it.[26]

There is no doubt that Tojo sympathised with ideas such as 'independence and sovereignty' and 'reciprocity' among the nations in Asia, which were proclaimed loudly at the Greater East Asia Conference. However, when he referred to the relationship between him and the leaders in the region, he often said that he would bring them over to his side.[27] Undoubtedly 'political strategy', not the ideas of independence, sovereignty and reciprocity, was always on his mind.

Tojo continued to be concerned with Burma and India in terms of 'political strategy'. In addition, he anticipated that Burma would be a target of the enemy's counter-offensive. In January 1943, he was already suggesting sending reinforcements to Burma in order to prepare against the counter-offensive.[28] In May, the Axis forces in North Africa virtually ended their war with the fall of Tunis. The Allied powers could now afford to send some of the naval and air forces stationed there to the Indian Ocean. Japan was therefore worried about the enemy's advance on the Andaman and the Nicobar Islands to the south of Burma, and since Burma was the south-western bastion of the 'Absolute Defence Sphere', now had to consider in earnest its defence, even if it was the sub-theatre of war in comparison with the Pacific. This is the origin of the tragic Imphal campaign.

It is not necessary here to explain in detail the Imphal operation.[29] The tragedy resulted from the faulty plan. The objective of the operation, which was primarily the offensive defence of Burma to forestall the enemy's attack, was not pursued consistently. Mutaguchi Ren'ya, who took charge of the operation as the Commander of the 15th Army, pursued an extraordinary objective of advancing into India, beyond defending Burma. In 1942, immediately after the victories in the first stage, it might have been a reasonable strategic option to advance into India in order to defeat Britain. But in late 1943 the war situation would not allow such a bold operation.

The faultiness of the plan was reflected most clearly in the underestimation of the enemy's strength and in the misjudging of the logistical problems. In July 1943, Sanada Joichiro, the Chief of the Operation Division, was impressed by 'the reckless aggressiveness' of the 15th Army's plan, when he heard the report of the IGHQ staff officer who had been sent to Burma.[30] At last, however, the IGHQ approved the Imphal operation in early January 1944. Tojo raised the following five points when he was informed of its approval:[31]

1 Is it possible to tackle the new situation when the British forces land on the Bay of Bengal in the south of Burma?
2 Is it necessary to reinforce the troops as the result of capturing Imphal? And is it against our advantage in terms of defending Burma?
3 Is the ground operation hindered by our inferior air strength?
4 Can the logistics catch up with the progress of the operation?
5 Is the operational plan of the 15th Army sound?

As usual, his questions made sense and hit the mark. Tojo instructed his subordinate to remind the General Staff of the five points he had raised.

However, as we now know well, 15th Army Commander Mutaguchi rushed recklessly to Imphal, in spite of the warnings and apprehensions of Tokyo. In mid-March 1944, when the Imphal operation had just started, Tojo, who had been Chief of the General Staff since February, observed, 'The first objective in Burma is to cut the route from India through northern Burma to China, and the second one is not to give the enemy the operational line to attack Thailand along the south-western coast of Burma.'[32] This observation hit the mark, too. He understood the strategic situation in Burma accurately. The objective of the operation for him was to forestall the enemy forces before they began full-scale attacks in northern and south-western Burma, and to defend Burma as a whole.

The Imphal operation appeared to work well at the start. In Japan, where the reports of defeat arrived successively from the battlefields in the Pacific, news of the victory in Burma looked like a gleam of hope. Tojo reported to the Emperor about the progress of the operation. 'We will achieve the objective before the rainy season which begins in mid-May, defeat the enemy in northern Burma and thoroughly cut the route from India to China.'[33] This was too optimistic a prospect, far removed from the realities of the battlefield. At the end of April the strength of the 15th Army had declined to 40 per cent, approaching the limits. Moreover, the rainy season began earlier than usual. It could not be denied that the operation had failed.

Finally, the IGHQ noticed that the operation had gone unfavourably, and in late April sent Hata Hikosaburo, the Vice-Chief of the General Staff, to Burma. When he returned to Tokyo, in mid-May, one of his staff asked him to report that the operation was absolutely a failure. But Hata reported to Tojo in a rather roundabout way that the prospects for the operation were extremely difficult. When Tojo heard Hata's report, he reproached Hata, and said, 'You can never tell until a battle is over. Don't be so faint-hearted.'

However, this was not what Tojo really meant. Since most of the senior officers in the War Ministry and the General Staff participated in the meeting where Hata reported, Tojo might have been worried that the Army's leadership would fall into defeatism. When Tojo talked with a few officers including Hata in another room after the meeting, he was perplexed, holding his head in his hands, and said, 'This is an awkward situation!'[34]

It is a more serious matter that Tojo did not order the operation to be stopped even after he knew the predicament in Burma. Hata also did not insist on stopping it. Hata thought that it was reasonable for the 15th Army to request the halting of the operation, since it was started at the 15th Army's request.[35] Tojo seemed to think in the same way as Hata. The decision to stop it was delayed, the victims on the battlefield increased and the defence of Burma as a whole was on the verge of crisis. As a result of the delay, Tojo allowed the situation in Burma to worsen. This was not because he lacked all sense of responsibility, but rather, it was because he lacked a clear strategic vision. Holding the post of Chief of the General Staff, Tojo could have played a leading role in making military strategy, but he would not take the initiative. Why did he not do so? It is because he did not have any strategic vision. Without this, he could do nothing but argue with empty rhetoric against 'faint-heartedness', or simply waited for his subordinates to present new ideas.

Conclusion

After the war, Sato Kenryo described Tojo's character as follows:[36]

> Mr. Tojo was never a dictator, nor had the makings of it. He was a man of circumspection, obstinate and tenacious, too. He had great power to persist in his own opinions and to execute them, never listened to anyone other than those in charge. His mental vision was rather narrow, so he looked like a dictator. And since he had wide jurisdiction, he seemed to be a dictator more and more, and people took it for granted that he was a dictator.
>
> But his mind had a weak side. He was always menaced by his own responsibilities since he had a too strong sense of responsibility. Certainly even true and vigorous dictators are sometimes menaced by their responsibilities, and not a few of them turn to gods and buddhas for help. Mr Tojo turned to the Emperor for help.

Sato's description of Tojo is very relevant and expresses his character skilfully. Tojo naturally had an unquestioning loyalty to the Emperor. His loyalty was so appreciated that he was appointed Prime Minister. And as a leader in war-time, he tried to bear the pressure of directing the war, with more intense loyalty to the Emperor.[37]

Tojo was a hard worker. He avoided as far as possible useless interviews with politicians and sake parties in the evening, but read reports and memoranda till late every night. His secretary would bring him a box full of reports and papers every evening, and he would have looked at all of them by the next morning and would return some of them with written instructions on important matters. When he was War Minister, not yet appointed Prime Minister, it was said that he could hold concurrently the various posts of Bureau Chiefs in the War Ministry since he knew everything about the bureaux. Nishiura guessed that Tojo could handle simultaneously even the duties of the various Section Chiefs, let alone those of the Bureau Chiefs.[38] It is understandable from this episode that afterwards he could concurrently and efficiently handle the posts of Prime Minister, War Minister and Chief of General Staff. He listened attentively to the reports of his subordinates, looked over all the papers brought to him, and gave proper instructions on matters requiring his approval. This was appreciated as Tojo's style of doing business.

Tojo took notes of every report by subordinates, and filed them according to his own classification in a special cabinet. He had three kinds of pocketbooks that were filled with his notes.[39] That is why he had to work late every night, and why business went smoothly. His questions and suggestions at the Liaison Conferences were based on these efforts and studies.

As stated above, Tojo's judgements on military affairs were objective in most cases, or at least not wide of the mark. His knowledge and insight as a professional soldier were reflected in these. But there are criticisms of him. He did not insist that his judgements were accepted. Although he offered proper questions and suggestions, he did not take the initiative in devising grand strategy. He did not present a strategic vision for victory. In this, he was probably constrained by institutional factors. In addition, his character undoubtedly influenced his conduct as a leader in wartime. Professor Ito Takashi, who edited Tojo's sayings and doings, pointed out one of the characteristics of his behaviour and thought patterns: 'Tojo emphasised repeatedly that it was imperative as an urgent task to win the war, but presented no notion of how the war would be won.'[40]

Perhaps Tojo did not have such a notion or vision. He said of himself to his entourage in wartime:

> People are making comments on me as a politician. But I very much hate being called a politician. I do not mind being called a tactician. I am never a politician. I am just applying the strategic method I learned from my experience in the Army.[41]

Obviously, his 'strategic method' was not a strategic vision at all. What he meant by 'strategic method' seemed to be responding flexibly and efficiently to the changes of situation on the battlefield. It was not creating the change of situation himself.

In peacetime, a practical leader like Tojo might be able to display his leadership even if he lacked a strategic vision. But lack of such a vision proved to be a serious defect of a war leader in the case of Tojo.

3

THE ARMY LEVEL OF COMMAND

General Sir William Slim and Fourteenth Army in Burma

Brian Bond

Slim is now widely regarded as one of Britain's best, if not *the* best Army commander in the Second World War. Some historians would place Montgomery in the top spot: he held the highest field command for longer than Slim and in three very different theatres – North Africa, Italy and North-West Europe. On the other hand Slim took command after a more clear-cut and humiliating defeat; his area of operations never enjoyed a high political priority; and conditions – in terms of geography, climate and logistics – were terrible almost beyond imagination.

What is not in dispute is that Slim was far more admired as a personality than Montgomery. The latter achieved renown with the British public as 'Monty', but Slim was affectionately and universally known as 'Uncle Bill', the soldiers' general *par excellence*. Slim has been described as a persuasive leader in contrast to Montgomery who was a dominant one.[1] Slim certainly looked every inch the military commander with his stocky figure, strong face and jutting jaw.[2] One has only to compare his appearance with the thin and rabbit-faced Percival, who surrendered at Singapore, to see what an advantage this was. But in Slim's case physical attributes truly reflected a sterling character. He combined the traits of essential modesty with complete self-confidence. He understood the very varied temperaments and qualities of his motley forces – Indian, British and colonial – and seems to have been equally popular with all of them.

In terms of Britain's grand strategy in the Second World War Burma can hardly even be rated as a secondary theatre; indeed its low priority along with Malaya goes far to account for the rapidity and severity of defeat in late 1941 and early 1942. Without US support, particularly in air transport, the reconquest of Burma could hardly have been contemplated, yet the United States' strategic interests were essentially concerned with logistical and military support to Chiang Kai-shek and not at all with the reconquest of British colonial territory. Consequently, it remained uncertain until late in the war that a north-to-south offensive through the most formidable terrain would even be attempted.

After the humiliating defeats in 1942 the Burma theatre received very little attention in the British media compared with Dunkirk, the blitz on British cities, the dramatic see-saw of the North African campaigns (in which General Rommel achieved a remarkable popularity) and the culminating advance from Normandy to the Baltic. Consequently, with some justice, British soldiers in India and Burma felt resentfully that they were a 'Forgotten Army' in a neglected theatre. The flamboyant, publicity-seeking Admiral Mountbatten attempted to alter this perception as Supreme Commander in the area from October 1943, but even so the names which established themselves with the British public were his own and that of Major-General Orde Wingate, the eccentric leader of the Chindits killed in an air crash in March 1944. Wingate was the equivalent of T. E. Lawrence in the First World War, his deeds and controversial personality overshadowing Slim's achievement as Lawrence's has done to Allenby's. As Duncan Anderson has pointed out in a perceptive essay, Slim only became one of 'Churchill's generals', that is to say known personally and respected by Britain's war leader, at the very end of the war.[3] Sir Basil Liddell Hart, for so long the chief authority on British generals' reputations, gives Slim only ten entries in the index of his *History of the Second World War*[4] against Mongomery's 56, but the discrepancy is in fact much greater because many of the latter's references are to several consecutive pages.

Even after the Second World War, Slim's had not become a household name in Britain despite his ascent to the head of the Army (Chief of the Imperial General Staff) in 1948. He had made only a fleeting appearance in the popular film *Burma Victory* and was, in effect, squeezed out of public attention by the more controversial characters and exploits of Mountbatten and Wingate. What changed these perceptions dramatically was the publication of Slim's memoirs of the campaign *Defeat into Victory* in 1956.[5] The book was very well written, explained complicated events lucidly and reflected the author's modesty and dignity in refraining from inflated claims and vendettas. Rarely in this genre has a commander admitted to making so many mistakes, some potentially disastrous, or been willing to attribute so much of his success to others. In some cultures, the qualities of modesty and understatement are not highly regarded, but in Britain they raised Slim's status to that of military hero. The Official History of the war in Burma subsequently portrayed Slim as mainly responsible for victory while displaying a surprising degree of hostility towards Wingate. Full-length studies by Sir Geoffrey Evans (one of his divisional commanders) in 1969, and Ronald Lewin (an excellent historian) in 1976, both praised him almost unreservedly. In Lewin's judgement he had achieved the lofty status of Sun Tzu's 'Heaven-born captain' or ever-victorious general.[6]

There is a lesson here for all successful generals, and perhaps also for students of military history; namely that reputations may be made and certainly enhanced by well-written and well-timed memoirs. Consider, for example, the influence of those of U. S. Grant, T. E. Lawrence, Montgomery and Guderian. Ronald Lewin revealed that Slim had published widely in the inter-war years under a pen name, and loved writing only a little less than soldiering. Lewin also revealed that Slim (again like Grant) had brilliantly worn the 'mask of command'; that is, so far from being callous or nerveless he was in reality sensitive, self-critical and prone to doubt. As his former chief engineer told Lewin: 'Of all his many attributes I never cease to admire his calmness and courtesy when the strain . . . for long periods on end must have been wellnigh unbearable. His imperturbability did not stem from insensitivity, but rather from a superhuman self-discipline.'[7]

Although Slim's professional career as a commander culminated in a brilliant and deserved victory in 1944–45, he had served a long and by no means entirely successful apprenticeship. Moreover, like most successful military leaders, he had enjoyed a fair measure of good fortune.

William Slim's origins differed sharply from those of most British generals of that era. His social background was lower-middle-class, his father being an unsuccessful Birmingham ironmonger. 'Bill' Slim, born in 1891, won a scholarship to attend the local grammar school, but his prospects seemed limited; despite his early interest in military history and soldiering, Sandhurst and an officer's career were not available to him for financial and social reasons. He was briefly an elementary schoolteacher in a Birmingham slum district, and then a junior clerk in a metal-tubing firm. Though never a student at the local university, he had been allowed to enlist in the Officer Training Corps (OTC) so that on the outbreak of war in 1914 he was one of the thousands of keen but strictly amateur soldiers who were given temporary commissions. His humble origins and tough work experience proved very useful in his natural rapport with ordinary soldiers.[8]

Bill Slim had a 'good war' in 1914–18 in that he displayed leadership qualities in brief but intense periods of active service (with the Royal Warwicks) in Gallipoli and Mesopotamia. He was severely wounded in both campaigns and sent to convalesce as a junior officer on the staff in India, where he transferred to the Army, because there he could live on his pay without a private income. He had witnessed the terrible results of incompetent administration and poor staff work, especially at Gallipoli; and would henceforth give these unglamorous aspects of soldiering a high priority. Also in Mesopotamia, he gained valuable insight into the causes of low morale and how they could be overcome.

In the inter-war period (1919–39) Slim was clearly picked out as a 'high flyer' among Indian Army officers. Selected for the Staff College at Quetta he passed out top to be rewarded by the 'plum' posting as the Indian Army representative on the instructing staff at the (British) Staff College at Camberley.

This was followed by an even more prestigious assignment to attend the course at the Imperial Defence College for the most promising officers of the three services, along with representatives from the Commonwealth and the Foreign Office. Slim had proved himself a good student and a good teacher with a sharp analytical mind and a firm grasp of military realities. He was not, however, noted as a theorist or an innovator. Regimental service with 6th and 7th Gurkhas provided experience with first-class soldiers and close friendships with several officers who would later become his senior subordinates in Burma.[9]

Even with these credentials few could have predicted Slim's eventual achievements in 1939. He had at last been promoted lieutenant-colonel but had been outdistanced in terms of age by several rivals in the Indian Army. This relatively slow progress was to prove a blessing in disguise. Britain's habitual military unpreparedness in peacetime entails that it is usually disastrous to be a senior commander at the start of war. Many of Slim's peers came to grief in the ferocious onslaughts by German and Japanese forces in 1940 and 1941, whereas he spent the first 30 months of the Second World War in military 'backwaters' fighting the far less formidable Italians and Vichy French.

Slim's first experience of leadership in combat, commanding 10th Indian Brigade against the Italians on the Sudan–Ethiopian border in November 1940 might well have been his last. His task was to recapture the fort of Gallabat just inside Sudan, followed by the strongly held nearby fortress of Metemma across the frontier. Gallabat was quickly taken but at high cost. Nine of Slim's 12 tanks were knocked out by mines or boulders and their crews were then shot up by his Garwhali troops who mistook them for Italians. With his own slender air cover destroyed, Italian bombers and fighters pounded Slim's troops inside the fort, and the Essex regiment, recently inserted into his brigade under protest, panicked and fled. This was a cause of lasting bitterness on the part of Essex regimental officers towards Slim, whom they accused of poor planning, and helps to account for General Noel Irwin's unsuccessful attempt to get Slim removed from his corps command in Burma in 1943.[10] After consulting his senior officers Slim decided not to risk an attack on Metemma, but instead withdrew to a ridge which dominated Gallabat and effectively deterred an Italian advance. Slim took full responsibility for this failure which was made to seem even worse when it emerged that the Metemma garrison was about to surrender when he called off the attack. He resolved in the future to be bolder and more aggressive.[11]

That he got another chance was due to a series of accidents resulting in his unexpected promotion to command of 10th Indian Division in Iraq in May 1941. In an operation in which poor communications and horrendous supply problems posed more difficulties than the Vichy French colonial garrison, Slim devised a risky two-pronged assault to capture Dier-ez-zor in eastern Syria. A minor crisis occurred when the column making a wide sweep through the desert ran out of fuel. The timetable went awry but Slim held his nerve and movement was resumed by draining the tanks of vehicles on the line of communications. The attack was completely successful; Slim's self-confidence was strengthened; and he gained further command experience in the ensuing occupation of Persia.[12]

At the beginning of 1942 Slim's future as a senior commander was far from assured, particularly as General Auchinleck regarded him as no more than a 'competent second division player'. Thanks, however, to the lobbying of two of Slim's former colleagues in 6th Gurkhas (Major-Generals Scott and Cowan), in March 1942 Slim was appointed Commander of Burma Corps under the newly arrived Army Commander-in-Chief, General Sir Harold Alexander.

This proved to be Slim's golden opportunity, but initially it looked more likely to signal an ignominious end to his career. With Malaya, Singapore and southern Burma already lost, the rapidly advancing Japanese ground forces, fortified by dominance in the air, looked unstoppable. These operations had already destroyed the career of one officer, General Tom Hutton, an able Chief of Staff suddenly handed the poisoned chalice of field command in a chaotic retreat, and would soon account for others, including Brigadier John Smyth VC, who was held responsible for the disaster at the Sittang bridge.

The difficulties facing Slim would have overwhelmed a less robust commander. His own retreating forces were defeated and demoralised whereas his formidable opponent held the initiative and his morale was sky-high. Slim's two divisions were poorly trained and ill co-ordinated; they were road-bound and, as yet, had no answer to Japanese tactics of rapid cross-country movements to set up road blocks in their rear.

At the time, Alexander received most of the credit for what was depicted as a heroic retreat in appalling conditions, but the historical verdict on his role is now very critical.[13] Alexander allowed operations to drift and never gave his corps commander clear directives. Was Slim to counter-attack, retain as much territory as possible, including the oilfields in central Burma, or was he to strive to keep Burma Corps intact? Slim's instinct was to attack, but his only realistic hope was to draw his two divisions together. This proved impossible because Alexander repeatedly allowed units to be detached to support the Chinese in eastern Burma. After several offensives had failed or been called off, Alexander finally, on 25 April, issued the order for a withdrawal to India.

Although in *Defeat into Victory* Slim would blame himself for several operational errors, and even admit to being indecisive at a critical point when Japanese infiltration seemed to have cut off the retreat across the Chindwin, the withdrawal may still be regarded as a remarkable achievement.[14] At Monywa and Shwegyin, for example, Slim extricated his forces from impending disaster with great skill and he displayed impressive powers of leadership combined with improvisation in narrowly winning the race, against not only the enemy but also the monsoon, in bringing his exhausted, hungry and disease-ridden troops to the comparative safety of Assam. Burma Corps' fighting soldiers retained their discipline and their morale, but these combat units were preceded across the frontier by an indisciplined mob, mostly of Indians from the line of communications. No longer in organised units, and having deserted their officers, wrote Slim, 'they banded together in gangs, looting, robbing, and not infrequently murdering the unfortunate villagers on their route'.[15]

Slim never lost heart or hope in these darkest days. He remained outwardly calm and confident, using every opportunity to make personal contact with his troops; skilfully varying his approach to achieve the biggest impact on different groups and nationalities. In effect he prevented a military disaster by sustaining morale. In Lewin's elegant summary, 'He did well in manoeuvring his divisions, but he did better in making them the partners of his spirit.'[16] In similar vein, Sir Geoffrey Evans suggests that the officers and troops of Burma Corps knew that Slim had done all that was humanly possible; hence the rousing farewell they gave him when he relinquished his command at Imphal. Evans even goes so far as to suggest that 'this was Slim's greatest test as a commander in the field and that the two months' operation was his finest contribution to eventual victory over the Japanese'.[17]

Slim's first step in turning defeat into victory was a cool analysis of the reasons for failure in 1942. Most of them lay beyond his own responsibility. There was, first, a terrible lack of preparation because no one in higher authority had expected an invasion of Burma. Consequently the ground forces provided to defend the country were utterly inadequate: two hastily assembled and inexperienced divisions, one of them equipped for desert warfare. The total elimination of the weak Allied air force was a crippling blow, but Slim believed the eventual key to victory would come from outfighting the enemy, soldier for soldier, on the ground. For Slim, however, the most distressing aspect had been the contrast between Allied generalship and the enemy's. The Japanese leadership had been confident, 'bold to the point of foolhardiness, and so aggressive that never for one day did they lose the initiative'. Poor Allied intelligence contributed to Japanese tactical dominance. Wide turning movements or 'hooks' through the jungle to set up road blocks on the Allied line of communications succeeded again and again, thereby breeding an inferiority complex.[18]

Slim reserved the severest criticism, though surely too harshly, for himself. He had repeatedly tried and failed to pass to the offensive and regain the initiative. He had not realised how the enemy, so formidable so long as they were allowed to advance, could be thrown into confusion by the unexpected. He resolved to profit from the lessons of a bitter defeat and to act more boldly in future.

Between mid-1942 and the end of 1943 Slim laid the theoretical and practical foundations for eventual victory. His principles and methods, profoundly relevant to the theme of military leadership, can only be briefly discussed here. Among the tactical lessons he inculcated were the following: the individual soldier must learn to move and fight in the jungle; he must conquer the tendency to panic when enemy parties infiltrated behind his lines and believe that it was the Japanese who were 'surrounded'; there should rarely be frontal attacks and never frontal attacks on narrow fronts; tanks could be used in almost any type of country except swamp; and, perhaps most important, 'there are no non-combatants in jungle warfare' – even medical units must be prepared to defend themselves.[19]

He goes on to describe in fascinating detail the steps he took to improve the health of his soldiers, including a more varied diet, better hospital facilities, tough measures to combat jungle diseases (especially malaria) and the air evacuation of serious casualties. He analysed the complex factor of morale into three elements: spiritual, intellectual and material. He had no doubts about the justice of the Allied cause; believed firmly that ordinary soldiers and not just officers would respond to reasoned appeals regarding leadership and eventual victory; and understood that soldiers would appreciate most of all that living conditions, equipment and weapons must all be as good as possible. An innovation which did a great deal to boost morale was the publication of a theatre newspaper – *Seac*. Mountbatten rather than Slim probably deserves the chief credit for enlisting the services as editor of a brilliant young journalist, Frank Owen, but the latter gave the venture his full support.[20]

Although other armies have made remarkable recoveries from humiliating defeat, Duncan Anderson is surely right to stress the magnitude of Slim's achievement in 1942 and 1943.[21] The theatre of war could hardly have been more forbidding: several hundred miles of virtually roadless jungle-clad mountains, swamped for half the year by the monsoon rains. The Army's role there was uncertain and hence its priority was very low for supplies and manpower. Perhaps most problematic of all were the mixed composition and poor motivation of Slim's forces: British, Indians, Gurkhas, East and West Africans, few of whom can have been deeply committed to the restoration of British control over Burma. Finally, after their experiences in 1941–42, Slim's motley forces were in awe of the supposedly unbeatable Japanese.

Slim's first step in raising morale was to convince his army that the Japanese were not supermen and could be defeated. He set up a realistic and demanding training schedule in India in which units were sent into the jungle for weeks at a time. The tactical emphasis was placed on attack rather than defence. The second, complementary step, was the great amount of time Slim devoted to travelling long distances to talk to his troops; in putting over his aims and ideals in simple language which all could understand. This was a rare attribute among British generals in either of the World Wars and is perhaps attributable to Slim's relatively humble origins, lack of higher education, and early experience in the Birmingham metal industry. Through this combination of practical reforms, improved training and doctrine and his genius for personal communication, Slim had revitalised Fourteenth Army and given it – something rarely evident at that level – a proud sense of identity. His universal nickname of 'Uncle Bill' encapsulates the magnitude of his achievement.

Although Slim had stamped his personality and ideas on Fourteenth Army he had no say on Allied grand strategy and only a limited influence on operational strategy within the Burma theatre. As an Army commander his chief responsibilities were to maintain relations with allies (notably Stilwell) and senior subordinates (notably Wingate) and, above all , to make a few vital decisions about the location and timing of offensives, responding to enemy initiatives and the summoning of reinforcements.

As Supreme Allied Commander in the theatre, from October 1943, Mountbatten deserves praise for shielding Slim from a good deal of the political interference from London which posed such a problem for field commanders nearer home and in the Middle East.[22] As the senior land forces commander under Mountbatten, General Sir George Giffard also worked well with Slim, but difficulties arose when he was replaced, in November 1944, by Sir Oliver Leese, who had taken over from Montgomery as Commander of Eighth Army in Italy.

Much can be gathered about Slim's personality and leadership qualities from his relations with the US commander in the China–Burma–India theatre, General 'Vinegar Joe' Stilwell. To describe the latter as 'prickly' would be a weak understatement. He appeared to despise British senior officers – whom he called 'limeys' – and refused point-blank to serve under Giffard. Yet he *was* willing to accept Slim's operational control. Thus was a 'military nonsense' created whereby Slim would bypass Giffard to report directly to Mountbatten regarding Stilwell's US and Chinese forces. This illogical command set-up worked because Stilwell respected Slim and generally agreed with him on broad objectives. When difficulties arose, Slim's sensible solution was to fly to Stilwell's headquarters and discuss matters face to face. Stilwell insisted that this unusual command arrangement was not to be made public, and it did not survive his own removal from the theatre of operations.[23]

Though they differed sharply in other ways, Slim's daily routine as an Army commander closely resembled Montgomery's. He would rise at 6.30 a.m. and spend the mornings studying the latest news and holding meetings with senior staff officers, air commanders and Allied representatives. He would leave his office at about three, read a novel for an hour, go for a walk in the cool of early evening with one of his staff, dine at 7.30, talk in the mess until 9.30, visit the operations room for a final look at the latest reports, and be in bed by ten. Between then and 6.30 he would only permit being roused for a real crisis. His belief in the need for leisure to think and for unbroken sleep to regain energy was surely wise. Generals who are terribly busy all day and half the night, he remarked, 'wear out not only their subordinates but themselves'.[24] They then lack the reserves of mental and physical vigour to deal with a real emergency. It is evident that neither Slim nor Montgomery could have tolerated for long Churchill's propensity for keeping senior officers at his beck and call until the early hours. Alanbrooke somehow endured this routine but at the cost of tremendous strain.

Slim's staff organisation is also worthy of comment. He never adopted the 'Chief of Staff' system in which the senior staff officer serves as the mouthpiece of the commander to other staff officers and heads of support services. Slim preferred the older method of dealing directly with his principal staff officers because he felt it was essential to project his own personality. Interestingly, Slim made his senior staff officer the major-general in charge of administration because he believed that in a theatre like Burma administrative considerations would loom larger than strategical and tactical issues.

It was Mountbatten who appointed Slim to the command of the newly created Fourteenth Army soon after he arrived at SEAC headquarters, and his continuing support was crucial throughout the campaign. Slim was glad to hear Mountbatten announce that there would be no more retreats and that, in event of a crisis, he would somehow find enough transport aircraft to maintain supplies. The two commanders also agreed that air transport would make it feasible to maintain operations through the monsoon, but this was only accepted reluctantly by some senior officers, including General Giffard. Giffard, for this and other reasons, lost the confidence of both Mountbatten and Slim. However, relations between Mountbatten and Slim remained amicable despite the former's monstrous egotism and uninhibited showmanship. According to Lady Slim, when Mountbatten visited Slim on his deathbed the latter was heard to remark, 'We did it together, old boy.'[25]

An important attribute of a military leader is to select loyal and efficient subordinates and keep them together as a unified team. Slim did not believe in taking all his staff with him when he was promoted, but he 'inherited' several former Gurkha colleagues, such as Scott and Cowan, when he arrived in Burma, and his selection of both staff officers and field commanders was admirable. Needless to say, a heroic defence leading on to an advance to victory breeds confidence and enhances a sense of team spirit, but it can hardly be doubted that in 1944–45 Slim had gathered, and was supported by, a first-class team including Stopford, Messervy (an import from Eighth Army), Rees, Cowan, Roberts, Evans and Christison. Among his outstanding staff officers were Snelling, Hasted, Steve Irwin and Lethbridge.

Slim had not selected Wingate and, like all the latter's nominal superiors, had difficulty in controlling him. In *Defeat Into Victory*, Slim leaves the impression that, although he clashed with Wingate on some specific issues and had to overrule him, he broadly accepted Wingates ideas and admired his imaginative schemes and dynamic leadership.

However, in his fine biography of Slim, Ronald Lewin revealed that the Commander of Fourteenth Army had deliberately muted his animosity towards Wingate the man and his objections to his policies. Slim had remarked in a private letter that Wingate had been 'deliberately untruthful in some of his statements and most disloyal in passing such statements behind the backs of some commanders to others'.[26]

Wingate's numerous supporters continue to argue that Slim was unfair to their hero,[27] but the latter's standpoint as the Army commander is easily understood. Wingate's ambitions for expanding his long-range penetration forces were tantamount to giving them the main role in the reconquest of Burma with the remainder of Fourteenth Army in effect assigned to a supporting role. Slim could not tolerate this challenge to his authority, not only because he believed that the diversion of air transport and supplies would undermine his own plan, but also because he was opposed to 'special forces' in principle, being confident that ordinary divisions could be adapted, if necessary, for airborne operations behind enemy lines. Slim's post-war judgement was to deny that the contribution of the Chindits was either great in effect or commensurate with the forces they absorbed.

In retrospect, Fourteenth Army's victory in the attritional battle for the Imphal plain and its lifeline to India, followed by the brilliant offensive operations which carried it across the Irrawaddy, has an air of dramatic inevitability. But at the time the sequence of events and their timing were far from clear. Slim, in particular, was largely forced to wait upon others' decisions, whether they were Mountbatten's or Kawabe's. Slim, like most of the strategic planners, would have preferred the main Allied offensive thrust to take the form of an amphibious landing somewhere on the coast of south-west Burma, with his Army and Stilwell's forces playing only the supporting roles in an overland north-to-south advance.[28] But at the end of 1943 it seemed extremely doubtful that the necessary shipping would be made available to Mountbatten. Consequently, Slim's strategic aim could only be the very general one of the total destruction of the Japanese forces in Burma.

In the first half of 1944 Slim's abilities as a defensive general were put to their severest test. In February a Japanese offensive in the Arakan took the British by surprise but this time there was no panic or hasty retreat. The decisive action was fought around the 'administrative box' where rear-area troops dug in and waited for air supplies. Two fresh British divisions advanced from the north to inflict a crushing defeat on the Japanese, who lost more than 5,000 men. This, however, was a secondary front to which Slim had committed four divisions to check what was essentially a Japanese diversion. Intelligence reports had suggested for several weeks that the main blow would fall on the Imphal front which was now thinly defended.

Slim's strategy was to draw the enemy into a battle of attrition on the Imphal plain where IV Corps would have an advantage in artillery and armour and where the Japanese lines of communications would be dangerously stretched. But Slim faced a critical command decision regarding the concentration of IV Corps' main component of two divisions each spread out and widely separated from the other in the mountains to the south. A premature withdrawal would suggest to the Japanese high command that their plans had been anticipated and might well prevent a decisive battle from taking place beyond the attackers' supply capacity. On the other hand if the order to withdraw was given too late the exposed British divisions might well be cut off and defeated piecemeal.

Slim was soon seen to have erred on the side of caution in giving orders to withdraw, because the Japanese offensive began on 4 March rather than on

15 March as he expected. His cardinal error, as he admitted in *Defeat into Victory*, was to leave this critical decision in the hands of IV Corps Commander Scoones since he, Slim, was better placed to judge the whole situation. The outcome was nearly disastrous.[29] Seventy miles south of Imphal, 17th Indian Division was, indeed, cut off on the Tiddim road but managed to escape after fierce fighting and with help from the Imphal garrison. Fifty miles east of Imphal, 20th Indian Division, though not cut off, also had to fight its way back along the Tamu road.

Slim also admits to what was potentially an even more serious error of judgement, namely that the attacker would not be able to supply more than a brigade group in a thrust towards Kohima and Dimapur, whereas it became clear that the whole Japanese 31st Division was committed in a daring long-range infiltration. Kohima was defended only by a small improvised garrison, Dimapur initially by none at all. Allied operations might just have survived the temporary loss of Kohima, but that of the base and railhead at Dimapur would have been disastrous. Its loss would have delayed the relief of Imphal, exposed British communications and airfields through the Brahmaputra valley and cut off Stilwell's operations with the Chinese on the Ledo front.

Although Sir Geoffrey Evans, later to serve as one of Slim's divisional commanders in the reconquest of Burma, has argued that the latter was too self-critical about this episode in *Defeat into Victory*,[30] the fact remains that the situation was only saved in the latter half of March by the remarkable airlift of 5th Indian Division with its infantry, guns, jeeps and mules from the Arakan to the Imphal plain. While 2nd British Division moved in more slowly by rail, the airlift of 5th Indian Division was only made possible by Mountbatten's unilateral decision to divert US transport aircraft from the China front. As the crisis in the Imphal plain developed, Slim was also unintentionally aided by the Japanese 31st Division's commander, Sato, who persisted in attacking at Imphal rather than pressing on to capture the more valuable prize of Dimapur.[31]

During these critical weeks of March and April 1944 Slim's personal leadership qualities were most severely tested – and proven. The physical strain was enormous as he shuttled between his widespread commands in uncomfortable aircraft in order to keep abreast of events, not only at Imphal and Kohima, but also in the Arakan, on the Ledo front and regarding the Chindits, who were by now operating behind Japanese lines. The moral pressure was perhaps even greater in the need to appear calm and confident when things were going wrong. Slim succeeded in a military leader's greatest challenge: to take charge of a battle in which the enemy has seized the initiative and by a wise disposition of existing forces and reserves turn the crisis decisively to his own side's advantage.[32]

As the battle of attrition ground on relentlessly through April and into May, Slim's initial strategic errors were more than offset by those of his opponents. Sato's relentless battering at Kohima gave Slim the chance to save Dimapur while, on a higher level of command, Mutaguchi's unwillingness to admit defeat condemned his 15th Army to a horrific retreat through the monsoon largely without food or medical supplies.

Although Imphal was not finally relieved and communications with Dimapur reopened until 22 June, Slim was already confident of victory by mid-May when he realised that the enemy divisions could be effectively destroyed even before their retreat began. With characteristic generosity Slim stressed that his troops had really won the battle, but he insisted that his strategy had been basically sound. It had been 'to meet the Japanese on ground of our own choosing, with a better line of communications behind us than behind them, to concentrate against them superior forces drawn from Arakan and India, to wear them down, and, when they were exhausted, to turn and destroy them'.[33]

With Mountbatten's strong support, Slim's forces pressed on with their appalling slog towards the plains of central Burma through the monsoon.[34] Only two divisions of XXXIII Corps conducted this struggle against nature and the elements, 11th East African Division marching from Tamu eastwards to the Chindwin at Sittaung, while 5th Indian made for Kalewa along the equally difficult Tiddim road. Both divisions were obliged to depend largely on air supply, and losses were high from sickness and disease. At an early stage it was realised that engineering resources made it impossible to convert more than a single route (down the Kabaw valley from Tamu to Kalewa) into an all-weather road. The Tiddim road was simply left to collapse as the Indian Division advanced. By mid-December 1944 bridgeheads over the Chindwin had been secured and an impressive Bailey bridge completed at Kalewa.

Logistics were therefore the critical factor as Slim and his staff began to plan for a decisive battle in central Burma. Staff studies suggested that the maximum force supportable beyond the Chindwin – some 400 miles from the railhead at Dimapur and 200 from the most advanced air-supply bases – was four complete divisions with two additional infantry and two tank brigades. Allowing for Japanese commitments in Arakan and on Stilwell's front, it seemed likely the enemy would deploy at least five divisions and numerous miscellaneous troops on the central front. Superior Allied air power would clearly be a great asset, but Fourteenth's Army's skill in more open mobile warfare remained to be tested.

Slim had assumed that the newly appointed Japanese commander, Kimura, would concentrate his forces for the decisive battle in the central plain north of the Irrawaddy. However Slim's two corps, XXXIII (under Stopford) and IV (under Messervy), made such rapid progress that suspicions grew that he had made a huge strategic miscalculation. Air reconnaissance showed that Japanese movement was eastward across the Irrawaddy and not towards the advancing British forces. Other intelligence sources confirmed this suspicion: the enemy had no intention of making a stand within the Irrawaddy loop, but were going to defend Mandalay with the great river barrier in front of them. Slim had been on the verge of pouring his army into a vast cul-de-sac where he would have been faced with a frontal attack to cross a wide and fiercely defended river before he could even begin the battle for Mandalay. Slim believed that this change in Japanese strategy was due to Kimura's generalship being more flexible and imaginative than his predecessor's (Kawabe's), but it later emerged that the high command in Tokyo had decided that southern Burma must be held at all costs; the territory within the Irrawaddy loop was consequently deemed to be expendable.[35]

Slim's drastic change of plan (from code name 'Capital' to 'Extended Capital') to deal with this unforeseen crisis was surely his most impressive achievement as a strategist.

The new plan, thoroughly discussed by Slim's staff but not cleared in advance with either ALFSEA or SEAC (i.e., Leese or Mountbatten), was described by Ronald Lewin as 'the most subtle, audacious and complex operation of his whole career'.[36] The essential feature of the plan was to convince the enemy that the main attack, by a whole corps, would be made across the Irrawaddy with Mandalay as its objective. Meanwhile, under cover of this deception, the other corps would approach the river further south using minor tracks to conceal its advance, cross at weakly held points and then drive all-out for Meiktila, the nerve centre of Japanese communications in central Burma. The second phase of this risky plan was to race for the sea to capture a port in southern Burma, Rangoon or perhaps Moulmein, before the advent of the monsoon in mid-May played havoc with Slim's already tenuous communications. Mountbatten would later describe Slim's master-stroke as 'a bold plan, relying for its fulfilment on secrecy, on speed and taking great administrative risks'.[37]

No matter how brilliant the plan, all would depend on the details of its implementation, which can only be summarised here. On 12 February 1945, 19th Division supported by XXXIII Corps, began a series of fiercely contested crossings of the Irrawaddy which confirmed the Japanese assumption that this was the main attack directed against Mandalay. Meanwhile in January and early February, IV Corps had moved some hundred miles to the west and marched south-east down little-known tracks to Pakokku nearly 50 miles south of the main Japanese defences. Its offensive, begun the day after XXXIII Corps' against Mandalay, did not go entirely smoothly or as planned,[38] but by 4 March its armoured spearhead had entered Meiktila. Kimura was taken by surprise, but once the threat to Meiktila became clear he rushed forces south to save this vital road and rail centre. The Japanese put up their usual fanatical defence of both cities, but Mandalay was finally captured on 21 March and Meiktila a week later. After only a few days to rest and regroup, Slim's forces raced south in an attempt to reach Rangoon before the monsoon broke. Unfortunately for Slim, the monsoon arrived unusually early on 2 May while his leading units were still some 50 miles north of the city. But this hold-up, though an irritating anti-climax, soon proved unimportant because an amphibious landing by XV Corps launched from Arakan captured Rangoon virtually unopposed.

The converging British forces linked up on 6 May so, to all intents and purposes, ending the campaign for the reconquest of Burma.

There was a scarcely credible episode in the days immediately following the victorious advance to Rangoon summarised above. General Sir Oliver Leese, Slim's immediate superior as Commander-in-Chief Land Forces South-East Asia, (ALFSEA), informed Slim that he was relieving him of his command and giving him the humdrum role of 'mopping up' Japanese resistance in Burma while his successor planned the invasion of Malaya. Slim apparently took this astonishing blow calmly and did not openly protest nor, as far as is known, did he lobby privately either in the theatre or with the authorities at home. But he did remark to one of his officers, referring to Irwin's earlier attempt to have him dismissed in 1943: 'Don't worry my boy, this happened to me once before and I bloody well took the job of the chap who sacked me. I'll bloody well do it again.'[39] This proved prophetic. There was a general upsurge of opinion in Slim's favour throughout the Fourteenth Army and in India; Alanbrooke (the CIGS) intervened decisively; and Leese was indeed dismissed to be succeeded, after a brief interval, by Slim.

In conclusion, Bill Slim was a tough and successful Army commander who was also a charming and humane character. In this he was unlike most victorious generals, and in possessing so many admirable traits we may feel he was 'almost too good to be true'. Yet this unlikely combination of positive qualities, professional and private, has never been seriously questioned, let alone 'debunked'. He was a professionally well-educated and highly trained soldier, a master of staff work and administration. He chose his subordinates wisely and gave them loyal support. He enjoyed good relations with most of his superior commanders (notably Mountbatten), and co-operated well with allies, including the notoriously prickly Stilwell. He was a good listener and adopted a relaxed, almost democratic, style in discussing plans with his staff, while leaving no uncertainty as to who bore the final responsibility for decisions. Slim was completely free from snobbery or self-importance; he excelled at winning the trust of all ranks and various nationalities – including even Chinese generals! One of his most important assets was his ability to speak to his troops in plain language which they could understand, explaining not only practical, soldierly matters but also dealing with more delicate issues of ideas and idealism. This gift for personal communication brought a rich reward in terms of loyalty and affection and goes far to account for his remarkable achievement in creating an *esprit de corps* at Army level.[40]

Slim's judgement as a strategist was sometimes faulty, but he never tried to gloss over his mistakes or put the blame on others. Nor, unlike Montgomery, did he later claim that his plans had all worked out perfectly in combat. He possessed the requisite physical robustness and the moral strength to overcome crises without showing the doubts and anxieties which privately assailed him.

In operational terms Slim demonstrated his leadership qualities in three very different circumstances: first, in sustaining morale and discipline through a devastating retreat which would have destroyed a lesser commander; secondly, by maintaining his strategic aim through a long defensive battle of attrition where the outcome hung in the balance; finally in orchestrating a victorious advance through most difficult terrain against a formidable enemy, crowning the operation with a daring, improvised plan which succeeded brilliantly.

Slim may have to share with Montgomery the accolade of 'the finest (British) general the Second World War produced',[41] but, as a sterling character on whom would-be military leaders might aim to model themselves, he was in a class of his own.

4

LEADERSHIP IN JAPAN'S PLANNING FOR WAR AGAINST BRITAIN

Kanji Akagi

Introduction

The Second World War saw what were, at root, unrelated wars in Asia and in Europe being fought as one. This situation developed in one fell swoop with Japan's attack on Pearl Harbor in 1941, but had arisen from the clarified delineation of allies and enemies that accompanied the Tripartite Pact in 1940, the 1941 attack by Germany on the Soviet Union, and the start of British and US aid to the Soviet Union. In addition, the United States both as a bulwark of the Western hemisphere and as a global power with Asian possessions in imminent danger, had become unable to remain an idle spectator in view of the possibility of the fall of Britain, which had already resulted in US assistance to Britain and *de facto* participation in the war in the Atlantic.

Two assumptions contributed to the drawing up of Japan's war plan. These were a belief in German invincibility and in the likelihood of British surrender. These assumptions had been shared by Japanese military policy makers consistently from the success of the May 1940 German blitzkrieg onward. The resolution to go to war was reached in the period from September to December 1941. In particular, participants at the Imperial Conference of 5 November deemed eventual war with Britain, the United States, the Netherlands and the Chungking regime to be unavoidable. The basic strategy for the war, the 'Plan for the Successful Conclusion of Hostilities with Great Britain, the United States, the Netherlands and the Chungking regime', was decided upon by the Imperial Headquarters–Government Liaison Conference (hereafter called the Liaison Conference) of 15 November.

Research on the pre-war situation shows that both the Japanese government and the supreme command recognised the high probability that any such conflict would be long and drawn out, and that Japan did not possess the necessary strategic material capabilities to sustain such a war or the means, therefore, to bring about a US surrender by force. For example, at the Liaison Conference,

Admiral Nagano Osami, Chief of Naval Staff, stated that Japan did not have the means to invade the United States in order to force the enemy to lose the will to fight and to surrender,[1] a view that was also shared by the Army. However, the form that the war would take was not agreed upon; 'a protracted war', implying a perception of invincibility, and 'a short and decisive war', anticipating a draw, being envisaged by the Army and Navy respectively.

Unable to force the enemy's surrender directly, Japan's war plan envisaged indirectly ending the war based on the two assumptions mentioned above. The plan emphasised the aim, firstly, of bringing about the British surrender and, subsequently, capitalising on this to end the war with at least a military draw and a negotiated settlement with the United States. This constituted 'the British factor'. This chapter seeks to investigate the extent to which 'the British factor' was considered or ignored in Japan's war plan, operations, plan building and implementation.[2]

The logic of Japan's war plan

On 5 November 1941, the Liaison Conference determined that war with Britain, the United States, the Netherlands and the Chungking regime was unavoidable. However, preparations for the formulation of a general war plan, 'a general outline for a war against Britain, the United States, the Netherlands and the Chungking regime', had officially been under way among the Army, Navy and Foreign Office since August of that year. The final section of this general outline was extracted and became the final 'Plan for the Successful Conclusion of Hostilities with Great Britain, the United States, the Netherlands and the Chungking Regime' ('Tai Ei-Bei-Ran-Sho Senso Shumatsu Sokushin ni kansuru Fukuan', hereafter *Fukuan*), formally adopted by the Liaison Conference on 15 November.[3] This became Japan's basic strategy, and was generally accepted as such throughout the government and military leadership. It constituted the only codified war plan completed before the commencement of hostilities.[4]

Whilst, as discussed above, planners recognised that Japan itself did not possess the means to make the United States surrender, it was not perceived that this would necessarily mean Japan's defeat. The minutes of discussions and related policy documents from the deliberations of the numerous Liaison Conference meetings held can be aptly summarized as follows.

> Victory was certain in short-term operations, and, if specific conditions held, the conflict could be brought to a military stalemate. However, longer-range prospects were uncertain. With regards to a long-term conflict, the upper echelons of the Army and Navy were on the whole pessimistic.[5]

Stipulations of the *Fukuan* were as follows:[6]

Course of action:
To 'rapidly overrun allied centres of resistance in the Far East area, thus gaining a position of strategic self-sufficiency, proceeding to further active measures to topple the Chiang government and defeat Britain, with German and Italian support, thereby undermining US morale'.

Main Objective 1
'The Empire will by means of force rapidly overrun allied centres of resistance in East Asia and the Western Pacific, thereby gaining a position of strategic advantage, secure important resource-rich regions and main lines of communication in preparation for long-term self-sufficiency, and by any necessary and appropriate means lure the main force of the US Navy into decisive battle and destroy it.'

Mention in the plan of the strategic option of aiming for a protracted war, and the strategic option of seeking a military deadlock by means of a short and decisive war deserves further discussion. The complications that would emerge after the completion of the initial phase of operations were already evident. If compelling the United States' surrender was impossible, the pursuit of a stalemate by the roundabout means of causing the United States to lose the will to continue fighting was the only remaining option. The most effective method for executing this indirect policy was deemed to be the defeat of Britain.

Imperial Headquarters pointed out the comparative weakness of Japan's Western Front, and the importance of attacking Britain and the Chungking government. It also emphasised the scenario of ending the war by destroying American morale and will to continue the war by means of the defeat of Britain. In addition, it suggested the following three methods of bringing about a British surrender in alliance with Germany and Italy: firstly, cutting off the links connecting Britain with India and Australia by strategic means as well as commerce raiding; secondly, hastening the independence of Burma, and using this to stimulate Indian independence; thirdly, in concert with the advance of Germany and Italy into the Near East, North Africa and the Suez region, beginning operations to advance into West Asia. Furthermore, planners also concluded that strengthening naval blockades against Britain, with Germany undertaking operations to occupy mainland Britain when the situation allowed, would force Britain to capitulate.[7] Only the first and second of these methods involved independent action by Japan, and any direct effects could be anticipated only with full reliance on Germany. Other than this, there was some support for the isolation of Australia and the United States in the plan, but, even if this involved Japan, it was not related to the policy of seeking the surrender of Britain.

While not directly concerning Japan, there were also some means by which Japan's co-operation could assist in the war being fought by Germany and Italy in Europe against Britain, and thus ensure a British surrender. Firstly, in concert with Germany's advance into the Caucasus, the Middle East and North Africa, Japan could advance into West Asia and India, and in co-operation with Germany and Italy threaten the sphere of influence of the British Empire in Asia. Secondly, Japan could mediate a separate peace between the Soviet Union and Germany, thereby alleviating the burden of the Soviet operations on Germany and enabling Berlin's war effort to be focused exclusively against Britain. To the same end and in accordance with German requests, Japan could invade the Soviet Union Far East, and indeed Germany consistently sought Japan's entry into the war against the USSR. However, because the maintenance of a neutral relationship between Japan and the Soviet Union was considered a condition indispensable to the success of the southern operations, for a long time Japan was unresponsive. Further, whilst the army and foreign ministries pursued the option of employing Japan's good offices in mediating a peace between the Soviet Union and Germany, throughout the war Germany maintained its position of seeking Japan's entry into the war, and the Soviet Union similarly was unresponsive to the possibility of concluding a peace with Germany. The biggest problem was that for a long time Japan did not comprehend the nature of the war between the Soviet Union and Germany.[8]

The international state of affairs itself was not as Japan had hoped, and shifts in the situation of the war in Europe did not unfold according to Japan's anticipations. Germany's military strategy exhibited a sudden turnaround as Hitler abandoned the idea of invading the British Isles in autumn 1940 and instead attacked the Soviet Union in June 1941. The war in the Soviet Union proceeded satisfactorily at first, until reaching an impasse with the coming of winter. Nonetheless, the Japanese Army continued to anticipate and trust in a German victory. When, in the spring of 1942, Germany reopened the offensive against the Soviet Union, the Army believed that the overthrow of the Soviet Union and then Britain would likely soon follow.

The 'Army and Navy Southern Operations Central Agreement' of 5 November 1941 sanctioned operations to launch simultaneous surprise attacks on the Philippines and British Malaya, as well as approving operations from both west and east to reach south as far as the Netherlands East Indies. The aim of the southern operation was, as seen before, the destruction and conquest of Allied centres of resistance in East Asia, and its main targets were the Philippines, Guam, Hong Kong, British Malaya, Burma and Netherlands East Indies (Java, Sumatra, Borneo, Celebes). However, the Central Agreement of Southern Operations did not include a plan of campaign to follow after the completion of the first-phase operations.[9]

The attack on Pearl Harbor at the start of the war was brought about, despite considerable opposition within the Naval Staff owing to the risky nature of the operation, by the strong resolution of the Commander-in-Chief of the

Combined Fleet, Admiral Yamamoto Isoroku. It was also accepted by the Army from the point of view of its significance in protecting the left strategic flank of the southern operation against US interference. As a result of the Pearl Harbor raid, Japan succeeded in its aim of destroying the main surface force of the US Navy in the Pacific.[10]

Complications surrounding the 'outline of the conduct of war' and second-phase operations

Japan declared war on Britain, the Netherlands and the United States on 8 December 1941. The first 100 days of the war saw astonishing advances by the Japanese armed forces, and the first phase of the war was marked by good progress. Hong Kong was conquered on 25 December, Manila was taken on 3 January 1942, Singapore fell on 15 February, Rangoon on 8 March and Java was occupied on 9 March causing the government of Netherlands East Indies to flee to Australia. By March 1942, the southern resource-rich regions had been occupied by Japan.

In the *Fukuan*, three scenarios had been suggested as possible opportunities for ending the war: firstly, when the main objective of the southern operations had been reached; secondly, during the main stage of the operations against China, especially upon the surrender of Chiang Kai-shek; and, thirdly, at the time a good opportunity presented itself with regard to the changing situation in Europe, for example, the end of the war between Germany and the Soviet Union, or following the success of measures against India.[11] It would appear that by March 1942 the first opportunity envisaged in this plan had arrived. However, the question remains of whether, after the initial phase was completed, the war plan was drawn up according to the grand strategy represented by the *Fukuan*. In other words, whether the 'westward' strategic policy that lent importance to 'the British factor' had been realised requires investigation.

On 7 March 1942, the Liaison Conference decided on the 'General Outline of the Future Conduct of the War' from among the studies being conducted for the next stage of operations. The *Fukuan* had been by nature decisively influenced by Germany's tendencies, but Germany did not show the slightest inclination to act according to Japan's wishes and consistently sought Japan's entry into the war with the Soviet Union, which sealed the fate of this plan as a means to end the war. The 'General Outline' had been formed in the victorious mood that had accompanied the success of the early stages of the war, but the outline that succeeded it was devised when the prospects for a British surrender had become bleak. In common with the previous plan, the new outline failed to resolve differences in strategy over whether a protracted war or a short decisive war was the objective, and was therefore ambiguous. In other words, the aims of 'continuously expanding on military achievements already accomplished and securing a condition of long-term political and military indestructibility whilst seizing any opportunity and actively taking measures in order to force a British surrender and make the United States lose the will to fight' were set down simultaneously. The order of priority of 'expanding on military achievements', 'securing a condition of long-term political and military indestructibility' and 'actively taking measures' was not made clear.[12] This bears witness to the first consideration being 'making it sound plausible', rather than ironing out the disparities in the strategy.[13]

In the end, Japan did not realise the goals of its westward strategy. The first reason for this is that, influenced by the huge gains in the early stages of the war, vigorous debate arose in the Army and Navy about the strategic direction of the advance in the period after the initial phase of the war. Critics argued that the rigorous conduct of war as set out in the *Fukuan* had not in fact occurred. Secondly, the idea of Japan advancing into western Asia (the westward strategic pincer strategy) had been based upon the perception of a golden opportunity to link up with the advance of the Afrika Korps then advancing into the Suez region and Egypt. Germany, however, locked in a desperate struggle with the Soviet Union, consistently sought Japan's entry into the war in the Soviet Union, while Japan continued to desire the maintenance of the status quo.

Professor Paul Kennedy has posited the hypothesis that, with Britain still reeling from a Japanese attack on Ceylon and south India in the spring of 1942, a move to seize Ceylon at this time could have been tremendously effective. In other words, the Japanese carrier fleet, based upon Ceylon, would not only have cut off India but also have dislocated the shipping route to the Persian Gulf and Egypt (and prevented the British build-up for El Alamein). Furthermore, a German–Japanese link-up in the Middle East was not so far-fetched at this stage in the war.[14]

Having investigated the possibilities for operations to follow on from the first phase of the war, on 15 April 1942 the navy decided on the second phase of operations. This war plan placed emphasis on active offensive operations, largely because of the strong influence of Admiral Yamamoto Isoroku, the Commander-in-Chief of the Combined Fleet. Various other ideas had emerged from the Army, but once, with the occupation of Java, the strategic aims of Japan's southern operations had been achieved, it was deemed that henceforth the basic aim should be to ensure preparedness for a protracted conflict.

The Navy's plan consisted of: firstly, destroying the British fleet in the Indian Ocean and, in concert with Germany and Italy's operations in West Asia, seizing Ceylon, cutting off the links between Britain and India, thus ensuring close collaboration with Germany and Italy; secondly, seizing Fiji, Samoa and New Caledonia, cutting off US air and sea lanes with Australia – the so-called FS operation – and, if possible, aiming for a future occupation of Australia; thirdly, occupying Midway and destroying operational bases there and seizing the Aleutian Islands in order to prevent a surprise enemy attack on the Japanese mainland and block the execution of any US offensive plans; fourthly, occupying the areas surrounding Hawaii (Johnston Island, Palmyra Island, etc.) to force a showdown with the US fleet. Should conditions allow, the plan also advocated seizing Hawaii in co-operation with the Army.[15]

The Navy's investigation of possible operations revealed a lack of mutual understanding between the Naval Staff in Tokyo and the Combined Fleet Headquarters, which had been exacerbated by the successes of the early stages of the war.[16] Further, even within the Combined Fleet Headquarters, there existed inconsistencies between the ideas of the Commander-in-Chief, Admiral Yamamoto, and those at the staff officer level. The Naval Staff sought the disruption of the lines of communication between Australia and the United States and argued for operations in Fiji and Samoa, while the Headquarters of the Combined Fleet, from Chief of Staff Rear-Admiral Ugaki Matome downward, saw the grand strategy in terms of the westward strategy and were studying the possibility of operations in the Indian Ocean and of the seizure of Ceylon, on the one hand, while elaborating on a plan to pursue the destruction of the US fleet in the Pacific, on the other.[17] The Naval Staff had a strong interest in the Pacific front against the United States, and, while not opposed to the concept of the occupation of Ceylon, were not overly supportive of such an operation.[18]

In March 1942, Admiral Yamamoto declared at the staff meeting that:

> the adoption of a defensive posture in anticipation of a protracted war is, for me as Commander-in-Chief of the Combined Fleet, impossible. The Navy must necessarily take an offensive posture and be dealing severe blows to the enemy at some location. The military power of the enemy is five if not ten times ours. Against this all we can possibly do is to continue to inflict ferocious assaults on the enemy where it hurts him. Resting on our laurels with our accomplishments and thinking that we can take on an invincible posture is no more than imbecility.[19]

The Army was perplexed at the Navy's apparently aggressive attempt to ignore the grand strategy represented by the *Fukuan*, and particularly concerned at the proposed eastward operation against the United States. Chief of the Operations Division of the General Staff Lieutenant-General Tanaka Shin'ichi wrote several memos expressing his concerns that 'the conduct of the war may be undergoing a shocking transformation', and that 'the offensive operations of the Navy in the Pacific will become host to the conduct of the war hereafter'.[20]

Regarding the plans held by the Army, Tanaka described in a memo at the end of December 1941 the targets of the operations of the time as being the war against the Soviet Union (Siberian offensive set for May 1942), the war with China (the operation against the Chungking regime set for mid-1942), the war to occupy India, the war against Britain in the Indian Ocean (to seize

Ceylon), and the West Asian operations. 'West Asia' refers to the Persia, Iraq and Arabia region, and therefore it can be assumed that this operation was envisaged as being carried out in concert with German North African offensives in Egypt.[21] In any case, the last three of these targets were faithful to the aims of the *Fukuan* in prioritising the overthrow of Britain. However, to Tanaka, the most important priority was launching a war against the Soviet Union from Manchuria in spring 1942.[22]

With the prospect of war against the Soviet Union already in sight, and with the operation to seize Burma having begun in January 1942, the Army opposed an advance into the Indian Ocean as premature. Another consideration was that a German advance into the Middle East would not be possible for some time yet, meaning that no strategic link-up could occur. At any rate, having occupied the southern resource-rich regions, the Army's basic policy of adopting a defensive posture in order to fight a protracted war was shaken somewhat by the clamour to exploit the unanticipated successes of the initial phase of the war, but it remained consistent.

In the light of this situation, operations in the Indian Ocean were limited and involved the Japanese Navy deploying two powerful task forces into the Bay of Bengal in April 1942 and attacking but not occupying Ceylon (5–9 April). For the rest of the war, operations in the Indian Ocean involved fewer than ten Japanese submarines, which together with German submarines went about disrupting maritime communication routes.[23]

The Navy's most shocking defeat of the second phase of the war occurred at the Battle of Midway in June 1942. The operation had aimed for the total destruction of US aircraft carriers, but Japan's fleet of aircraft carriers in fact suffered near-annihilation. It is thought that Yamamoto anticipated an opportunity to bring the war to a successful conclusion by seizing Hawaii, and to this end it was necessary for the Navy to secure a succession of victories against the US and British Navies. Yamamoto's vision of repeated, decisive victories broke down quickly at Midway and proved unrealistic. Needless to say, the Army's war plan of defending the southern resource-rich areas and waiting for German victory also lacked any prospect of success.[24]

Following the defeat at Midway and in concert with the advance of Germany and Italy in North Africa, the Navy decided to bring a reorganised Combined Fleet into operation in the Indian Ocean. After the fall of Tobruk on 21 June, Tanaka recommended the commencement of the invasion of Ceylon to Prime Minister Tojo Hideki on 26 June.[25] On 11 July , Chief of the Naval Staff Admiral Nagano reported to the Emperor that the Navy's plans of operation had changed.[26] The modifications anticipated the German destruction of the Atlantic sea lines of communication and the German offensive in North Africa, and sought to shift the focus of operations to the Indian Ocean. The idea was that Japan, combined with Italy and Germany, could then attack the Middle East from both strategic flanks, which, together with a blow to merchant shipping, would drive Britain to collapse. The military achievements desired of Germany were unchanged but the emphasis of operations was once again restored to the west. The Navy's plan of operations envisaged taking control of an enormous expanse of the Indian Ocean from Ceylon to the Chagos Archipelago, extending as far as Madagascar, by the use of submarines and main naval forces.[27]

Admiral Nagano also reported to the Emperor regarding the cancellation of the FS operation that was to have taken place following Midway. However, having the momentum to pursue further military gains as set down in the outline, the Navy continued offensive preparations based on severing the lines of communication between Australia and the United States. In order to defend the naval base at Truk Islands in the Western Pacific, the Navy had been marshalling its air arm at Rabaul on New Britain and constructing an air base on Guadalcanal in the Solomon Islands. The Naval Staff had already cancelled the FS operation and its ideas about operations in the Indian Ocean had been transformed, yet a minor operation originally related to the FS operation would proceed. The imprudent advance into Guadalcanal was to lead to a great failure.

In midsummer, the Combined Fleet concentrated units in Singapore for the Indian Ocean operation. However, on 7 August, the US First Marine Division counter-attacked Japanese units on Guadalcanal. The Japanese Navy responded to this promptly and vigorously, and the main fleet and air arm of the Japanese Navy was brought into the eastern Solomon Islands.[28] The fierce battles for Guadalcanal continued until February 1943 and forced Japan to cancel all the planned westward operations, not only the Indian Ocean operation but also the planned war with the Soviet Union and the operation against the Chungking regime, until the end of 1942.

The elimination of 'the British factor'

Japan lost hundreds of aircraft, along with much of their crews, as well as aircraft carriers, warships, merchant shipping, oil and weaponry in the battle of Guadalcanal. With the completion of the withdrawal from Guadalcanal on 7 February 1943, the General Staff and the Naval Staff began to examine possible directions for subsequent operations. On 27 February, when the Liaison Conference conducted a 'Review of the World Situation', the prevalent opinion was that a German attack on the British Isles, advance into West Asia and pursuit of peace with the Soviet Union were all impossible. Prime Minister Tojo expressed doubts about the policy of forcing Britain's surrender that had been assumed since the *Fukuan*. General Sugiyama Hajime, Chief of the General Staff, declared that the idea of first forcing Britain to surrender and then causing the United States to lose its will to fight should be changed. Instead, emphasis should be on effecting the collapse of American morale.[29] Unable to work out a detailed plan, the Liaison Conference confirmed a policy of 'securing a position of strategic invincibility', which meant the elimination of the concept of a short decisive war and confirmed the anticipation of a protracted war. This also meant the elimination of 'the British factor', whose assumptions had inclued a 'westward' grand strategy concept.

Having heard the conclusions of this conference, the Army General Staff Section for the Conduct of the War began to examine new ideas for ending the war. Among their range of options, they recognised that, 'in the case of the defeat of the Axis powers, Japan will have no choice but to submit to obey Britain and the United States'.[30]

Conclusion

Japan's war plan was dependent on two assumptions regarding the war in Europe: British surrender and German invincibility. At the start of the war, the Liaison Conference had sought the rapid conclusion of the conflict by means of occupying the southern resource-rich regions, using the materials to promote the cultivation of military power, and conducting a protracted war of endurance until a negotiated peace settlement could be achieved. The policy makers of the Imperial Headquarters and the government had in fact anticipated the war turning out to be long term. However, they did not foresee the large-scale consumption of modern weaponry, the repeated fierce battles fought on land, sea and in the air, in short, the actual conditions of long-term war. The hoped-for military deadlock, leading to the degradation of the war into a struggle of endurance, in which the enemy would lose the will to fight, and to the convening of a peace conference never materialised.

The main force behind this idea was the view of war of the Japanese policy makers at the time. The Chief of Operations Section of the Naval Staff at the opening of the war, Captain Tomioka Sadatoshi, recalled:

> Our plan for the war was from the beginning to obtain a balance of power by inflicting heavy damage on the enemy, and then seek a point of compromise and make peace leaving Japan enough room to rise up again. However, there had been no endorsement of the hope for making peace forthcoming. Nonetheless, at that time and considering the contemporaneous war in Europe, the top leadership thought that an opportunity for peace would emerge since Germany seemed to be winning and causing a favourable shift in the balance of power.[31]

The wishful thinking about the 'rapid emergence of peace due to balance of power considerations' resulted from the fact that, although Japanese policy makers talked of the prospect of a protracted total war, most remained unclear about what this would mean in practice. At the start of the war, most Japanese policy makers thought of the end of a war in terms of the classic example of the Russo-Japanese War of 1904–05 and had what Tomioka described as a 'limited war view', the main cause of this being the failure to grasp the essence of total war from studying the First World War.[32] Japanese policy makers thus could not prepare an adequate grand strategy for total war.[33] Within the framework of the Japanese policy makers' thinking at that time, they were not able to imagine a 'total war' that culminated in an 'unconditional surrender', which would become one of the war aims of the Allied powers.

A lack of leadership in the Japanese military establishment was distinctly evident in the planning of the second phase of war operations. This stemmed not only from disorientation produced by the surprising flush of victory in the first phase of operations but also from organisational weaknesses within the planning staff. No strong leadership emerged capable of addressing these problems. In the case of the Japanese Navy, for example, insufficient manpower in the Combined Fleet Headquarters and a lack of mutual understanding with the Naval Staff in Tokyo impeded the efficiency and responsiveness of war planning. There was no form of leadership within the Japanese Navy capable of instituting organisational reforms, such as the integration of Naval Staff Operations Division and Combined Fleet Headquarters. As such, the organisation of the Japanese Navy was allowed to remain unchanged since the Russo-Japanese War.

5

CRISIS OF COMMAND

Major-General Gordon Bennett and British military effectiveness in the Malayan Campaign, 1941–42

Carl Bridge

'I approve of all methods of attacking provided they are directed at the point where the enemy's army is weakest and where the terrain favours them least.'
Frederick the Great, *Instructions for his Generals* (1747)[1]

'Australians are not easily impressed by British generals . . .'
Field Marshal Earl Wavell[2]

On 15 February 1942, after 70 days of fighting down the length of the Malayan peninsula and on to the island of Singapore, a British Empire force of 130,000 men was defeated by a determined and brilliantly led Japanese force of 60,000. It was the greatest capitulation in British military history since the surrender of Yorktown in 1781 during the American War of Independence. Arguably, the fall of Singapore signalled the loss of the second British Empire as much as that of Yorktown did the first. From a British point of view: how much should the defeat should be attributed to relatively inferior military leadership and effectiveness and how much to factors beyond the commanders' control?

The reasons for the Singapore defeat are many. At the level of grand strategy, the British Empire had chosen to concentrate on the European war against Germany and Italy. In the Far East, Britain would rely on deterrence and on the likelihood of United States belligerency. This meant that only two capital ships, the *Prince of Wales* and *Repulse*, were sent to the Singapore naval base against a known potential Japanese force of seven times that strength. The two British capital ships were sunk from the air at the outset of the campaign; the United States fleet in Hawaii, which might have assisted, was severely disabled in the raid on Pearl Harbor.

Malaya was allocated only 158 outdated aircraft. The Japanese invasion force had 617 modern machines. (By the time the British rushed in 157 new aircraft it was too late.) The British had no tanks; the Japanese had 80 tanks. Of the British Imperial forces, many were untried and some virtually untrained. They eventually consisted of a British division, an Australian division, two Indian divisions, two detached Indian brigades and a mixed British and Malayan brigade.[3] The Japanese used two battle-hardened divisions from the China front and the elite Imperial Guards. Thus the Japanese had a fully trained and balanced force while the British did not.

The British lost command of the sea virtually immediately and command of the air soon afterwards. Yet, to the bitter end they refused to panic the civilian population by building new fixed defences. They thought it more important to keep the vital rubber production up for the war effort.[4] Tactically, the Japanese commander and air co-operation expert Lieutenant-General Tomoyuki Yamashita's rapidly moving 'blitzkrieg on bicycles',[5] supported by his aerial artillery, and with his land-based and seaborne outflanking movements, kept Lieutenant-General Arthur E. Percival's British forces constantly off-balance. The campaign became a race in which Japanese speed outstripped the British capacity to bring in sufficient reinforcements to stabilise the situation.

All of the above notwithstanding, might Percival have played his hand better and secured better results? The commander of 8 Australian Division, Major-General H. Gordon Bennett, certainly thought so. Controversially, Bennett deserted his post at the surrender and escaped to Australia in order to teach his countrymen the lessons of the campaign before they too faced invasion.[6]

I

In 1941 Gordon Bennett was Australia's most senior citizen soldier. His half-dozen superiors in the Australian Imperial Force (AIF) were all professional soldiers. He had fought in the Dardanelles as a young officer in 1915. Moving to the Western Front, he had risen by the end of 1916 to be the youngest brigadier-general in all of the British Empire forces at the age of 29. His hero and model was the Australian Supreme Commander and citizen soldier General Sir John Monash who had led the AIF when it formed a vital part of the spearhead which pierced the Hindenburg Line and assured the Allies of victory in 1918. Bennett was physically brave, wedded to efficiency, and an advocate of attack at almost any cost. Though an accountant in civilian life, he despised what he saw as the stodginess and carefulness of regular staff officers, Australian and especially British, and he worshipped the dash of the amateur. Red-haired and of slight build and medium height, he was short-tempered, arrogant and outspoken, with more than his fair share of 'the privileged irascibility of senior officers'.[7] He constantly reminded his political and military masters that, in his opinion, he was cut out to be the Monash of the new war if only they would recognise it.[8] Bennett was given command of 8 Division in September 1940 and proceeded with it to Malaya in February 1941.

Bennett's superior officer in Malaya was Percival.[9] Both men were 53 years old. Percival was a regular soldier who had risen to brigadier by 1918 almost as rapidly as had Bennett himself. Between the wars Percival had served in a variety of mostly staff posts. In 1936–37, he was the colonel in Malaya Command who prepared the latest British defence plan. This plan revolved around air defence, with the Army guarding the airfields of the peninsula, particularly those in the north close to the Thai border at Kota Bharu and Alor Star. He commanded a division in home defence early in the war and in May 1941 he was appointed General Officer Commanding Malaya. Air Chief Marshal Sir Robert Brooke-Popham, Commander-in-Chief Far East, was his immediate superior.

What manner of man was Percival? One journalist described him as 'a tall, thin person, whose most conspicuous characteristics were two protruding rabbit teeth'. Brooke-Popham's successor, Lieutenant-General Sir Henry Pownall, thought Percival a man with 'the knowledge but not the personality to carry through a tough fight' who was 'an uninspiring leader and rather gloomy'.[10] Percival was cautious, diplomatic and studious rather than a strong leader of men. In his 1944 book, without naming names, Bennett railed against men of Percival's type:

> In war, a good soldier must be physically aggressive and a good commander must be mentally aggressive . . . In the days of peace, . . . the genial, respectful 'yes man' was accepted and given promotion and appointments . . . Our staff colleges have become somewhat pedantic . . . The hall-mark 'p.s.c.' (passed staff college) had produced thinkers rather than fighters.
>
> As a result of all of this, we have seen in the senior ranks of the British Army too large a percentage of brilliant staff officers and too small a percentage of aggressive fighters.[11]

A page later he described Percival as having 'features which were 'intellectual rather than dynamic'.[12] In his private diaries Bennett was much more direct: 'he does not seem strong, rather the Yes man type. Listens a lot but says little . . . Weak and hesitant though brainy.' Most indiscreetly and undiplomatically, though entirely in character, Bennett let his opinion slip at a press conference in August 1941 when he pronounced Percival 'clever but weak'.[13]

Bennett also had a low opinion of Percival's subordinate corps and divisional commanders. He wrote to the Australian Minister for the Army after the campaign started that 'my impression is that senior Officers in command here are not fit for their jobs'. They lacked 'aggressive spirit' and suffered from what he called the 'retreat complex'. He was particularly critical of Lieutenant-

General Sir Lewis 'Piggy' Heath, Commanding Officer of III Indian Corps. Heath had a brilliant fighting record on the Indian frontier and in Abyssinia, but was too anxious to retreat in northern Malaya. 'I spoke frankly about my concern for withdrawals. Urged attack . . . Heath objected . . . Very negative mind . . . I lost my head . . . After conference had a heart to heart talk with Percival. Told him it was better to lose Heath than the War.' Elsewhere, Bennett noted that Heath 'had a stronger personality than Percival and generally managed to impose his will'.[14] After the war, Percival admitted that he gradually lost confidence in Heath: 'all the way down the peninsula, I had a feeling that unless I issued definitive orders as to how long such and such a position was to be held, I should find that it had been evacuated prematurely'.[15] Heath and Percival were at loggerheads over strategy throughout the campaign. However, Percival did not have the courage to dismiss him.

The quality of the junior officers and troops also left much to be desired. Bennett thought too many of 'the [British and Indian] unit commanders lacked an aggressive spirit' and had 'a predisposition to "stickiness"'. They quickly became 'tired and weak' and were 'lacking in determination and pugnacity', though he also admired some of the British Indian Army officers who were personally brave and set an example to their men.[16] Some of the late arrivals among the troops were 'only partially trained as soldiers and were novices on jungle warfare' and 'it was unfair to commit them to jungle fighting against an experienced and victorious Japanese Army'.[17] Percival later wrote more charitably that 'It was not that there was anything wrong with the raw material but simply that it was raw.' In a private letter Bennett wrote that they 'scattered like schoolgirls'.[18] Some of the Australians were not any better. Among the later reinforcements some had never fired a rifle and others were the sweepings of the jails.[19] Nevertheless, the bulk of 8 Division was well trained, as were the British regulars, such as the Argyll and Sutherland Highlanders, and some of the Indian and Malayan formations. The problem lay not so much in the troops or their training, in their numbers or in their equipment, but in how they were used by their commanders, and with what air and sea support they were given.

II

More important, perhaps, Percival's general strategy was flawed. His plan, which was a version of that which he had prepared in 1936–37, was to defend the airfields in the north and to set his troops across the main rail and road route south, which ran slightly inland down the west coast. These tasks were allocated to III Indian Corps on the west coast and to two detached Indian brigades on the east coast at Kota Bharu in the north and Kuantan on the central coast. The job of defending Johore in the south, the essential hinterland to Singapore island and its base, was allotted to 8 Australian Division. The defence of the island itself was entrusted to British regulars and Malayan volunteers. These dispositions, while they covered all contingencies, left Percival's forces widely dispersed and vulnerable to defeat in detail. The defence of the airfields was misguided. Why defend airfields when the aircraft were not available in sufficient numbers or quality to use them? Certainly the plans to destroy much of the invading force from the air came to little. It might have been better simply to destroy some of the airfields and free the defending troops and the aircraft for other work.[20]

Percival's conduct of operations, too, had its weaknesses. The initial Japanese landings just over the Thai border at Singora and Patani were expected. It was planned to send a strong column into Thailand in a pre-emptive strike at a critical stretch of road called the Ledge, the demolition of which would severely hamper Japanese progress. Rather than make an immediate decision in the early morning, Percival referred it up to Brooke-Popham and half a day was lost, which was just enough time to allow the Japanese troops to get to the Ledge first.[21] Once it became clear that Yamashita's main thrust was to be down the west coast, Percival was still overcautious. Rather than concentrating his forces to meet the thrust, Percival chose to leave them spread evenly in case another thrust of equal weight should come on the east coast. This was a luxury he could ill afford and was a major contributing factor to the losses at Slim River and later on the Johore Line. In fact the road network in the east was so poor and the jungle so thick that a land-based advance was bound to be slower than one on the west coast. By not switching his force to the critical point, Percival gave Yamashita's smaller force the opportunity of achieving local superiority, and Yamashita took it.[22] One perceptive commentator argues that Yamashita was playing a game of Go, concentrating on the sweeping and rapid encirclement of his enemy by means of command of territory. On the other hand, Percival was playing a game of chess, trying with great deliberation to displace his opponent's pieces from their squares and to deny his opponent the squares occupied by his own.[23]

Another factor in the failure of the northern defence, and one rarely mentioned, is Bennett's determination to keep 8 Australian Division together and with its own specific task. Well before the outbreak of hostilities Bennett had negotiated with Percival that the Australians would be responsible for the defence of Johore, and Percival was reluctant to break up the Australian formation, and therefore Bennett's command, in order to provide a stronger reserve in the north. Had an action such as the one the Australians later fought at Gemas been fought earlier in the campaign, at say Slim River, the Japanese might have been delayed sufficiently to bring in the reinforcements Percival so critically needed. Bennett's personal touchiness and *amour propre*, coupled with his national pride, got in the way of Percival's making best use of the AIF. Had Percival been a stronger commander he might have had his way with the moody and difficult Australian.[24]

The basic Japanese tactics were a variant of blitzkrieg. Yamashita would probe the British position, seek to pin the main force with artillery, mortar and air attack and then send in a number of outflanking forces. Once one of these managed to get alongside or behind the British lines, particularly if this was on the main trunk road, the British would have little option but to withdraw.[25] With command of the sea virtually from the outset, Yamashita had the capacity to leapfrog his men down the coast. This he did first and most spectacularly at Telok Anson. As a direct result of this wide sweep in behind his opponent, he forced Heath's corps into headlong retreat in the battle at Slim River.[26] Yamashita used the technique again and again. Had the *Prince of Wales* and *Repulse* and their air cover still been intact Yamashita would not have had this option.

III

In his 1944 apologia, Bennett makes much of his claim to having devised from first principles a method of countering the Japanese tactics:

> Throughout the Malayan campaign our [British] military leaders were searching for the answer to this form of attack. The Australians did find a satisfactory method of overcoming these tactics. This method was later applied with success in New Guinea.[27]

He described the method in another passage:

> To hold a static defensive position, with the limited visibility that exists in the jungle and the correspondingly limited range of our weapons, is not easy. The attackers can quickly feel their way round the flanks of the defences, their encircling movement passing within a few hundred yards of the defending troops. This the Japanese did. The only answer was to attack. Strong fighting patrols working well to the flanks of the position could have held up the advance of the outflanking enemy units and, supported by a bold counter-attack, could have destroyed these units.[28]

Elsewhere in the book he extolled the virtues of guerrilla raids, mobile reserves and actively seeking out and destroying the enemy and his assets wherever they might be found rather than defending patches of territory.

What Bennett did not say in his book, however, was that the key elements of the method had already been developed in the first weeks of fighting in the north by 2 Battalion Argyll and Sutherland Highlanders under their dynamic Commanding Officer Brigadier I. M. Stewart. Stewart's tactics involved fighting 'for the road off the road' and hitting the enemy hard with highly mobile infantry in encounter battles. They not only pre-dated Bennett's similar tactics by some time but they were more highly developed.[29] Nor did Bennett's book say that Stewart was evacuated from Singapore before it fell so that he could teach his methods to the Australians in Australia. Thus it was the Scot Stewart rather than their own countryman Bennett who sowed the seeds for later Australian tactical successes in Papua and New Guinea using jungle warfare methods.[30]

More interesting, however, from our point of view, particularly considering his acid criticisms of his fellows, is to examine just how effective a jungle commander Bennett himself proved to be in Malaya. To make this assessment we must look at the actions at Muar and Gemas on the Johore Line (in reality more an area than a line) in mid-January 1942 and the defence of the critical north-western sector in the battle for Singapore island in early February.

IV

In early January, General Sir Archibald Wavell arrived in the theatre as Commander-in-Chief of the newly created ABDA[31] Area, which included Malaya. Wavell immediately and decisively overruled Percival and placed the more aggressive Bennett rather than Heath in command in the critical western half of the Johore Line. Bennett's Westforce included his own 27 Australian Brigade, two depleted brigades of 9 Indian Division, and the newly arrived but raw 45 Indian Brigade. He quickly set about organising a defence based on holding the enemy in the Gemas–Segamat area of the trunk road. To cover his flank on the coast he deployed 45 Brigade spreading them in a thin screen from the ferry crossing and coast road at Muar for 23 miles inland along the Muar River. His main force he positioned at Gemas and Segamat, with the Australians in front and 9 Division further south. The central element of his plan was a large battalion-strength ambush, which he set at the Gemencheh River bridge near Gemas where the trunk road bottle-necked. He then planned a strong counter-punch. He gambled on the Japanese not sending forces of any strength by the coast route through Muar.[32]

The Gemas ambush was a great success. Some 700–800 Japanese on bicycles were caught within the killing ground. Half of them died and many more were wounded. In the words of the Australian official historian:

> The charge [under the bridge] hurled timber, bicycles and bodies skyward in a deadly blast. Almost simultaneously, [Captain] Duffy's three platoons hurled grenades among the enemy and swept them with fire from Bren guns, Tommy guns and rifles. The din was so great that when Duffy ordered artillery fire the artillery forward observation officer thought his battery's guns were [already] firing.[33]

Before Bennett could reap any more permanent dividend from this, however, he was severely checked at Muar. He had instructed each Indian battalion along the extended river front to station two companies on the northern bank, thereby making a dangerously thin coverage even thinner. To make matters worse, he had allocated only one artillery battery to the whole brigade. He had deliberately starved Muar of resources in order to strengthen Gemas. This was to badly underestimate Yamashita. Yamashita pushed across the Muar River and at the same time landed another force further south, repeating the successful Telok Anson manoeuvre. Rather than counter-punch, as planned, Bennett was forced to try to shore up his flank, as 45 Brigade disintegrated. (Bennett had asked for 22 Australian Brigade from the east coast for this task and Wavell had agreed, but Percival, perhaps fatefully, had overruled him.[34]) In the event, Bennett chose to feed his Australians into the area a battalion at a time and suffered defeat in detail. There was further disappointment when the British 53 Brigade in its first action collapsed in Bennett's rear at Batu Pahat. Though a courageous rearguard action was fought, including a fierce two-day battle at Bakri, Bennett was forced to retreat as Heath had been before him. When tested, Bennett had done little better than those he had criticised.[35] The consequences of Bennett's faulty disposition were drastic: with the breaching of the Johore Line the battle for the mainland was lost.

It is worth noting here in passing that the Gemas ambush was not the only one of its kind in the battle for Johore. There were other ambushes on the west coast at Bakri by the 2/29 Battalion AIF and at Labis by 22 Indian Brigade, and one on the east coast at Nithsdale near Mersing by 2/18 Battalion AIF. The method did catch on, but too late to have any substantial effect on the impetus of Yamashita's campaign.

V

Bennett's dispositions and decisions in his next and last test, the battle for Singapore island, were equally questionable. Again, he was not helped by Percival. Just as Percival chose to try to defend everything on the mainland – and in doing so defended nothing adequately – so he did on the island itself. Though both Wavell and Bennett thought Yamashita's main thrust would be on the north-western side of the Johore causeway, Percival thought it might yet come on the north-eastern, and divided his forces to meet both contingencies. Percival also withheld committing his only reserve brigade to the north-west until it was too late. 'I had learnt on exercises we held in England not to commit your reserve until you are quite certain you are dealing with the real thing', he later recalled.[36] Unfortunately for him, what he thought was a feint in the north-west proved to be the real thing. Percival's error of judgement was to prove very unfortunate for Bennett.

First, however, a word about fixed defences. Winston Churchill was horrified to discover at this time that Percival had neglected to build obstacles on the beaches and to erect fixed defences of all sorts on the landward side of the island overlooking the Johore Strait. Astonishingly, Percival had rejected timely plans by his chief engineer to remedy the situation. Percival's grounds were that such defences were 'bad for the morale of the troops and civilians'. Losing his temper, the engineer replied to Percival, 'Sir, it's going to be worse for morale if the Japanese start running all over the island'; and Wavell was later to say something similar to Percival on the same subject. (Bennett, too, for different reasons, had earlier rejected tank obstacles proffered by the same engineer, saying he would rather have his gunners 'stop and destroy tanks with anti-tank weapons'.) The psychologist Norman Dixon has commented of both Percival and Bennett in this regard that 'to erect defences would have been to admit to themselves the danger in which they stood' and such an admission was anathema to their military training.[37]

Now let us resume our analysis of Bennett's command decisions. In the north-western sector of the island, Bennett deployed his two AIF brigades – ironically, they were together for the first time in the campaign – to cover the mangrove swamps and jungle-covered headlands facing the strait. He placed 44 Indian Brigade and his Malayan Brigade behind the Australians. With these relatively meagre resources, he had to defend some 30 miles of coastline. Needless to say, there were large gaps between his dispositions, gaps through which the enemy could penetrate. Ranged against Bennett's force, on a ten-mile front of their own choosing, were two Japanese divisions. In the aerial bombardment beforehand, part of Bennett's mobile headquarters was hit, destroying some of his records. 'Anyhow', he wrote sanguinely, 'a little less paper in this war will improve matters'.[38]

During the Japanese landings, which came at night, communications became chaotic with the telephone lines cut in the bombardment, artillery was not adequately co-ordinated and air cover was sparse and unreliable. Tired and confused, each of Bennett's brigadiers committed a serious error. Brigadier Taylor, ordered to advance at dawn, actually did the opposite as he thought he was about to be cut off by infiltrating enemy troops. Consequently, a hole was left in the line near the Tengah airfield and it was lost. In the circumstances, Bennett, who had noted Taylor's confusion, perhaps should have gone forward to assess the situation for himself. Brigadier Maxwell, who had been responsible for the Gemas ambush, made an even graver mistake. Maxwell misinterpreted an order from Percival to form a final inner perimeter, an order which Bennett had paraphrased verbally and released prematurely. As a direct result, Maxwell retreated to a point which opened the way to the vital Bukit Timah heights – site of the island's two reservoirs. Further, Maxwell's withdrawal triggered a more general retreat and the loss of the whole Kranji–Jurong Line. In his book, Bennett blames Maxwell for this, but it was actually Bennett's fault for releasing Percival's order too early and too casually.[39]

These were Bennett's worst moments. Percival commented: 'Gordon Bennett was not quite so confident as he had been up-country. He had always been very certain that his Australians would never let the Japanese through and the penetration of his defences has upset him.'[40] It seems that by this stage Bennett had virtually lost interest in the battle. At one point Bennett was asked by his staff to determine three phases of a planned counter-attack. Quickly and without thinking he slashed three lines across his map with his pencil oblivious to the fact that his proposed start line was partly in enemy territory.[41] Even before the retreat to the island, Bennett was anticipating defeat and thinking of escaping to Australia at the surrender.

Bennett and two of his staff escaped the moment the capitulation terms came into force. When Bennett arrived in Australia, however, he was accused of desertion. He was not formally punished and was given further military posts, but he never commanded troops in action again.[42] He had failed to emulate his First World War hero Monash in battle. Further, the monumental lack of political judgement shown by his decision to escape was definitely not something of which the master would have approved. There is no avoiding the strong suspicion that Bennett's mind in the last weeks was more on escape than on commanding his troops.

Arguably, the game was up for the island's defence once the Johore airfields were lost. But it is also arguable that the island might have held out until adequate reinforcements arrived. Two battle-hardened AIF divisions from the Middle East and a British armoured brigade from the United Kingdom were steaming across the Indian Ocean for Sumatra and due at the end of February. Seen in this context Bennett's and Percival's mistakes in Johore and on the island finally sealed the fate of the campaign.

VI

Frederick the Great's maxim that the best general attacks where the enemy is weakest and the terrain suits him the least is deceptively simple. Finding this *schwerpunkt* (centre of gravity) of an enemy requires not only reading the terrain, and his deployments and capabilities, and getting inside the mind of your opponent, but understanding your own capabilities and those of your forces. The iron laws of logistics can be pushed to the limit but not exceeded without penalty. In the all-arms warfare which characterised the Second World War it was vital not only to have air and sea support but to make maximum use of it. In the Malayan campaign Percival had, to quote one scholar, 'a bad hand' and he 'played [it] badly'.[43] He failed to concentrate his forces and tried to be strong everywhere, which meant he was strong nowhere. He repeated this pattern right down the peninsula. This made it easier for Yamashita to gain the local ascendancy necessary to break through time after time. Percival was too weak to curb Heath's 'retreat complex' or to bring Bennett sufficiently to heel to have him release an AIF brigade for the north. Bennett, among others, saw the tactical solution for dealing with the Japanese method, but he was incapable of carrying it out with any consistency. More often than not, Bennett made the same mistakes he was criticising others for making.

Military effectiveness, however, is a relative matter. Yamashita and his 25 Army trained hard, were properly equipped, had appropriate, flexible and effective tactics and fought a brilliant campaign. Had Percival, Heath and

Bennett played their hands faultlessly, it is still very doubtful that they would have bought sufficient time to change the ultimate result. Yamashita had a window of opportunity when the naval and air balance was in his favour and before land reinforcements arrived and he utilised it to the full.[44] At the time and in the circumstances the British Empire's generals and forces in Malaya were no match.

6

GENERAL YAMASHITA AND HIS STYLE OF LEADERSHIP

The Malaya/Singapore campaign

Kyoichi Tachikawa

Ever since the days of the Pacific War, or the Great East Asian War (1941–45) in Japanese, General Yamashita Tomoyuki has been highly esteemed as a victorious general during the first stages of the war, the Imperial Army's equal to Admiral Yamamoto Isoroku. Yamashita's fame spread overseas. Arthur Swinson, for example, wrote that '[i]n many quarters it was being said that he was the finest general in the entire army, and indeed the finest Japan had ever known'.[1] His image as a hero was amplified by the story and accompanying documentary film footage of Yamashita forcing General Arthur Ernest Percival to answer just 'yes' or 'no' during the cease-fire talks at the Ford Automobile Works north-west of Singapore City on 15 February 1942. Yamashita's good build and his handsome face that we can see in photos or films of the period, combined with his nickname 'Tiger of Malaya', made him famous.

His aides viewed him in a slightly different light. Lieutenant-Colonel Tsuji Masanobu, chief staff officer in charge of operations, under Yamashita, Commander of the 25th Army during the Malaya/Singapore campaign, wrote that:

> having a brilliant career and grand face and build, he was a man whose dignity was perceived both by the Japanese and by the non-Japanese. Perhaps because he had long played an active part at the centre of the military administration, he was able to pay close attention to details, was bright despite his imposing appearance and, to a large extent, he had a political turn of mind.[2]

Major Kunitake Teruto, staff officer in charge of operations in the 25th Army, wrote that:

> from the general opinion, judging by his large build, I had supposed that he should be a general with a resolute character and a manly, broad

mind, and never be shaken by anything. But I had not known that he was also delicate and sensitive.[3]

Regarding his sharp demand of Percival, Yamashita himself explained afterwards that he was so irritated by his slow interpreter that he involuntarily shouted 'yes or no', though he had intended to tell the interpreter to simply ask 'yes or no?', and that the scene was overstated when it was reported.[4] His image in photos and films is certainly imposing, but the truth is that he consciously projected his chest in order not to be seen as weak.[5] Moreover, the nickname 'Tiger of Malaya' was nothing but a nuisance for him. There is a Japanese expression *tora ni naru*, literally meaning 'to become a tiger', but it is also a metaphor meaning 'to be drunk', and Yamashita detested the nickname for that reason.[6]

It can be assumed that we have looked at only one side of General Yamashita or that our image of the general is a creation of the wartime press. It is certain that his reputation as the victorious commander of the Malaya/Singapore campaign would not be ruined because the images did not match the facts. Commanders should be evaluated by their conduct of war.

However, the details of his conduct of operations in Malaya and Singapore have not been fully examined. What kind of decisions did he make? For what reasons? Did his decisions have an influence on the course of the campaign as a whole? Is it possible to identify any characteristics and any reflections of his personality in the decisions?

General Yamashita and the planning of the Malaya/Singapore campaign

Let us examine Yamashita's role in the process of operation planning. Strictly speaking, the Malaya/Singapore campaign consisted of two campaigns: one to conquer Malaya and the other to conquer Singapore. The respective plans were not simultaneously prepared.

According to Major Kunitake, who was in charge of planning both operations *par hasard*, the study for the operation to conquer Malaya was set in motion in the Operations Section, Headquarters of the General Staff, in August 1940, and its plan was formulated by the end of October 1941. During this period, Yamashita was Inspector-General of the Army Air Service and Director of its Headquarters, and then was appointed Commander of the newly created Kwantung Defence Army. Moreover, he visited Berlin and Rome, and then was stationed in Shinkyo (capital of Manchukuo), so he was out of Japan for almost half a year before the outbreak of the Pacific War. Thus he had nothing to do with the planning of the Malaya campaign, though the final plan was approved in his presence as Commander of the 25th Army in November 1941. In the planning process, therefore, Yamashita 'was unable to put in a single word'.[7]

On the other hand, the operation to conquer Singapore was planned on Yamashita's initiative. He thought that the Japanese occupation of Kuala Lumpur made it certain that the Malaya campaign would be successful. At the same time, he showed Tsuji 'his original idea how to capture Singapore' and let him 'work out the plan of operation carefully'. Yamashita intended 'to cross the [Johore] Strait, then to draw attention to the reservoirs at the nearby heights and to call on [the British] to surrender. If they would not surrender, we would then carry out a battle of annihilation using all our military strength.'[8] Based on this outline, Kunitake prepared the plan to attack Singapore during breaks in his busy staff duties connected with the rest of the Malaya campaign, advised by Tsuji and Colonel Iketani Hanjiro, Chief, Operations Section, 25th Army. Yamashita approved Kunitake's plan on 30 January, the day when the leading troops of the 25th Army advanced into Johore Bahru, though he amended it slightly just before its execution, as will be touched upon later. In any case, it is certain that Yamashita was intimately engaged in the planning of the Singapore campaign. However, there is one thing to be added. Yamashita's troops were to cross the Johore Strait west of the causeway. The crossing area had been virtually decided as part of the plan of the Malaya campaign in Tokyo before Yamashita was appointed Commander of the 25th Army.

By and large, the Malaya/Singapore campaign was completed as planned, though it took only 70 days, a whole month shorter than General Sugiyama Hajime, the Chief of the General Staff, predicted during the Imperial Conference on 5 November 1941.[9] 'The operation progressed very rapidly, while units arrived one by one in the long term . . .' and 'more than 100 orders were given to allow the newly-arrived troops to take up their positions . . . During the entire campaign, not more than four or five times did the Army issue major orders involving the basic operational policy or the divisions' tasks and objectives', wrote Kunitake.[10] Therefore, circumstances seldom gave Yamashita opportunities to make important decisions. This was true of the Malaya operations in particular.

Malaya campaign

The first situation requiring Yamashita to make a decision occurred just after his landing on Singora on 8 December 1941. The 25th Army had actually prepared two plans concerning subsequent action: Plan A, after its successful disembarkation, 'to advance rapidly on the line of the Perak River and defeat the British Army',[11] and 'Plan B, in case unfortunately the British Army advances to the landing points beforehand',[12] that is, 'in case it became difficult for the main force of the Army to disembark in the first phase of the operation, to let part of the troops advance into southern Thailand and postpone the landing of the main force until the arrival of air units'.[13] It is said that the Imperial General Headquarters had insisted on Plan B in order to secure a bridgehead, but the 25th Army had prepared for its further movements along the lines of Plan A. In accordance with the Army's original policy, Yamashita adopted Plan A to push through to the Perak River line and advance to Singapore.[14]

Though the fact that the British did not disrupt the Japanese landing was surely a prerequisite for his choice, Yamashita enumerated the other reasons for his decision: there were no good airfields around Singora that appeared operational within a short time for repair; he wanted to secure the bridges over the Perak River before they could be destroyed by the British Army; and the military situation would become disadvantageous for the Japanese if British reinforcements should arrive.[15] It is also notable that, when Yamashita saw the jungles and the rubber plantations around Saigon for the first time in November 1941, he felt that this tropical environment was 'suitable for constant offensive operations . . . only if communications were carefully maintained'.[16] Is it possible to say that this intuitive reaction encouraged him to order the storming of enemy strong points? Yamashita tends to be regarded as a thoughtful officer, but he also relied on a strong sense of intuition.[17]

In retrospect, his decision was appropriate. The British were forced to fight the advancing Japanese from still incomplete defensive positions. Facing the onrushing Japanese troops, the British positions were broken through one after another and their morale was rapidly declining. It can be said that this psychological collapse was a crucial factor of the Japanese success in this campaign.[18]

Though his view is now regarded as racist, Yamashita thought that 'if Indians are included in the [British] troops, it is easy to deal with them'.[19] He had estimated the British Indian Army as low both in morale and in strength.

The outline of large unit leadership, Tosuikoryo (Principles of Command),[20] drawn up in 1928 by General Staff Headquarters, Japanese Army, stated that:

> according to conditions, operations on interior lines are often effective, too. This is particularly true of operations against an enemy with superior strength but with inferior fighting spirit. Thus this kind of operation tends to become passive. If every attack and defeat is not fully completed, it will later prove harmful. Therefore, its conduct requires the skill to take advantage of the operational situation by especially brave and bold decision and manoeuvre.[21]

Though it is debatable whether the Malaya campaign can be recognised as an operation on interior lines, it can be said that Yamashita's decision to advance boldly – making the most of manoeuvre 'against an enemy with superior strength but with inferior fighting spirit' – was sound. However, it is notable that behind Yamashita's decision to attack was his underestimation of British military strength.

Yamashita's second important decision was the release of his reserve 56th Division. In the initial plan of the 25th Army, in addition to the Imperial Guards, the 5th and the 18th Divisions, the 56th was going to be committed as the strategic reserve, though it was then assembling in mainland Japan. On 24 December 1941, Lieutenant-General Aoki Shigemasa, the Vice-Chief of Staff, Southern Army, and Lieutenant-Colonel Arao Okikatsu, a staff officer, paid a visit to Yamashita's command which was then located at Taiping and asked Yamashita to divert some of the troops under the 25th Army to another area (the Philippines in this case). Guessing the difficulties of the Southern Army in the Philippines, Yamashita agreed to the diversion of the 56th Division and of several artillery units,[22] which were 'being considered as trump cards for the attack on Singapore',[23] though, judging from the speedy development of the campaign, the 56th Division was not necessarily needed and the division would probably not arrive in time for operations. Yet even though the advance troops had already reached the Perak River, the initial goal of Yamashita's Army, and the order to cross the river had been given to them, it can be imagined that this was not an easy decision to make only a couple of weeks after D-Day, and it was naturally impossible to foresee the future of the campaign.[24] Kunitake wrote in his memoirs that 'when we agreed with the Southern Army, we felt as if we had jumped off the deck of the Kiyomizu Temple'.[25] This is a rare example of an army releasing part of its strength during an operation.

It is, however, essential to indicate that, at the root of his decision, Yamashita incorrectly underestimated the British military strength in Malaya and Singapore. At the beginning of the campaign, he and his staff inferred that, on the British side, there could be 'at most 50,000 soldiers or at the low end about 30,000'.[26] But as we know now, the British had more than double the Japanese calculation.[27] This underestimation was not corrected until the fall of Singapore. As Tsuji justly wrote, 'ignorance is bliss'.[28] Even Yamashita himself remarked later that 'our battle in Malaya was successful because we took the enemy lightly'.[29]

Let us suppose that the 56th Division had remained in Yamashita's hand. When the division had finished assembling and had deployed with the forces of the 25th Army, Yamashita's Army would, of course, have been reinforced, but at the same time 50,000 soldiers in total would then have been crowded into Johore Bahru, the tip of the Malayan peninsula, which would have aggravated the shortage of ammunition and rations. Then it would have been even more difficult for the troops to have operated at peak efficiency. It would, of course, have been unlikely for the 56th Division to find itself in Singapore before its fall. From the viewpoint of the overall Japanese Southern theatre, even though future events could not be foreseen, it was better to use the 56th Division in another area.[30] In that sense, it may be said that Yamashita's decision to release the 56th Division was justified.

It goes without saying that Yamashita had an influence on the handling of the other divisions under his command, too. Regarding the 18th and the

Imperial Guards Divisions in particular, it is easy to trace the reflections of Yamashita's intentions.

Lieutenant-General Mutaguchi Renya, in whom Yamashita had confidence, was Commander of the 18th Division. Among the troops under Mutaguchi, only the Takumi Task Force made its landing on Kota Bharu on D-Day and continued to drive south. Mutaguchi's main force was kept waiting at Cam Ranh Bay for orders to go into action. The Southern Army devised two plans for this main force of the 18th Division to conduct amphibious assaults on the British defensive positions along the eastern coast of the Malayan peninsula (Operations Q and S). Accordingly, the 25th Army changed its operation plans at Alor Star on 17 December 1941. Later, in view of the speedy development of his campaign, Yamashita realised that no opposed landings by the 18th Division were so required. Needless to say, there was always considerable danger in such landings. Rather than letting Mutaguchi and his troops do too much along the eastern coast of the Malayan peninsula, Yamashita hoped that the elite unit would play an active part in conquering Singapore and wanted to maintain its strength for that purpose by avoiding major losses.[31] At the same time, giving priority to attack against the British ships at anchor in Singapore, the Japanese Navy objected to the opposed landings because they diverted scarce resources needed elsewhere. The Southern Army still tried to conduct the landings, but, mainly because the Navy never wavered on its position, finally gave way on their plans. Ultimately, without taking any risk, as Yamashita hoped, the main force of the 18th Division landed unopposed on Singora on 22 January 1942.[32]

Instead, Takumi Task Force bore the brunt of the fighting. Even before Operation Q was officially cancelled, on 15 December 1941, Yamashita ordered Major-General Takumi Hiroshi to seize Kuantan, the objective of Operation Q.[33] Not equipped with transport or river-crossing equipment needed for a long overland march, Takumi 'consented' to Yamashita's order.[34] Fortunately, Takumi completed his task, but it is debatable whether this operation of Takumi Task Force should be recognised as an effective employment of strength or as a temporary expedient.

It was remarkable how cordially Yamashita received Mutaguchi. When the 18th Division landed on Singora, the leading troops of the 25th Army had already reached Kuluan. It was approximately 1,000 kilometres from Singora to Kuluan. The 18th Division had to march overland, but it had fewer than 100 trucks, and the rail network was heavily burdened transporting military supplies and equipment. Then Yamashita ordered the Imperial Guards and the 5th Divisions to provide 400 trucks for the main force of the 18th Division to drive south.[35] This might be both Yamashita's regard for Mutaguchi, who had been kept waiting for a month after D-Day, and his intention not to tire the officers and soldiers before they attacked Singapore. Thanks to this additional transportation, the 18th Division reached Kuluan on 29 January.

Here is another episode concerning Yamashita and Mutaguchi during the Singapore campaign. To Mutaguchi, wounded by an enemy hand grenade after crossing the Johore Strait, Yamashita dispatched Major-General Manaki Takanobu, Vice-Chief of his General Staff, with a letter of appreciation and a bottle of wine to express his sympathy. Mutaguchi appreciated this kindness so deeply that he felt obliged to fight more desperately.[36] This is an example showing that Yamashita's warmhearted personality worked well to motivate his men.[37]

Yamashita had known Mutaguchi and Lieutenant-General Matsui Takuro, Commander of the 5th Division, for a long time, but he did not know Lieutenant-General Nishimura Takuma, Commander of the Imperial Guards Division, well. Though this might not be a decisive factor, several conflicts occurred between Yamashita's command and Nishimura's division during the campaign, and Yamashita lost confidence in Nishimura and his division, as will be touched upon later.

Tsuji commented in his memoirs that:

> as the Imperial Guards Division consisted of soldiers selected from all parts of the country, each soldier had a better capability than those of any other divisions but it is regrettable that they had experienced no actual fighting since the Russo-Japanese War had ended. As they had traditionally been disciplined to be overly refined as well as courteous, they not only tended to be unsuitable for field operations but their leaders tended to defy the Commander of the Army.[38]

Because of these characteristics, Yamashita needed to be careful not to make the Imperial Guards Division lose face. Moreover, three commanders of the divisions under Yamashita were all in the same class (the 22nd) at the Military Academy, so he had to be considerate enough to treat and work the three as equally as possible. That is why the Imperial Guards and the 5th Divisions advanced in parallel on the front during the second part of the Malaya campaign though the strength of the Imperial Guards was only half of the 5th's.

Once at least during the Malaya campaign, there was confusion concerning the conduct of operations. The occasion was the battle between the Japanese 5th Division and the British Indian Army on the Kampar heights just after crossing the Perak River (from 29 December 1941 to 2 January 1942). On the morning of 2 January 1942, when the 5th Division encountered difficulty breaking the British defensive stand there, Kunitake, a 25th Army staff officer dispatched to the division, was informed that Watanabe Task Force, a provisional unit of the 5th Division, sailing south along the western coast of the Malayan peninsula, had not landed on Selangor, its originally planned landing point, but had landed on Melintang in the south-west of Kampar. Thinking it proper to let Watanabe Task Force advance in the direction of Sungkai and of the Slim River line in the rear of the British Indian Army in Kampar to cut off its retreat, Kunitake was permitted to recommend this action to Watanabe Task Force by the Command of the 25th Army, through Tsuji, the Army's chief staff officer in charge of operations, who was also dispatched to the 5th Division. But early in the evening, informed that 'part of Watanabe Task Force had landed on Selangor', the Command of the 25th Army changed its policy again and issued the 5th Division a new order to let the task force go toward Selangor as planned initially. The Command chose to cut off the British retreat at the farther point from Kampar rather than at the nearer. General Matsui, Commander of the 5th Division, tried to obey this new order, but Tsuji was infuriated at this change of orders, because the 25th Army had switched an agreed plan for one that might place the front-line fighting units in an awkward position. Furthermore, this was done without consulting him, the chief staff officer in charge of operations. Once the 25th Army realised that only a small part of Watanabe Task Force had reached Selangor, the operation was executed as Kunitake had planned, but Tsuji was still so resentful that he demanded to be relieved of his duties.[39]

It is not clear to what extent Yamashita was engaged in these frequent changes of orders, but no order of the 25th Army could be issued without his approval. It can be said that the steps followed by Kunitake were appropriate. On the other hand, the next order of the 25th Army to the 5th Division was based on vague and unreliable intelligence and given without any consultation with Tsuji or Kunitake who were the staff officers in charge of operations. So the steps followed by the Command of the 25th Army were undoubtedly wrong and too rash. In fact, Yamashita was so impatient by nature that his family nicknamed him *sugu-sugu-san*,[40] which might be translated as 'Mr Quickly-Quickly'. It is possible that the battle of Kampar was an example of his conduct of operations influenced by this flaw in his character. However, Yamashita was cool-headed enough at the same time not to accept Tsuji's resignation and expected that the troublesome officer would calm down.

This was one of the few cases where mutual understanding between the Command of the 25th Army and the front-line units was critically lacking during the Malaya/Singapore campaign. As mentioned earlier, since Yamashita had seen the jungles and rubber plantations around Saigon for the first time and had felt it necessary to keep in careful contact with his units, he had tried keeping in touch with the front line by dispatching his staff officers there from time to time. This was an unusual way of working staff officers. That might well be because Yamashita had to get along with Tsuji, a salient staff officer. Tsuji himself admitted that:

> the way of doing things in the Command of the 25th Army was quite different from that of traditional higher commands whose dispatched staff officers had customarily visited the front-line commands after the fighting was over and arrived by car or by plane, with a bottle of sake as a souvenir from their army commander. But we staff officers [of the

25th Army] visited battlefields in turns, even when hard fighting was still under way.[41]

Yamashita continually let his staff officers, including their Chief and Vice-Chief of Staff, go back and forth between his Command and the front lines so that he could let them oversee operations and grasp tactical situations. In this way, Yamashita himself was able to understand the strength and the situation at the front.[42] As Kunitake wrote later, however, it is regrettable that such a contact was not done sufficiently well between Yamashita's Command and that of the Imperial Guards Division.[43]

The Singapore campaign

Now, let us examine Yamashita's leadership in the Singapore campaign. As mentioned before, regarding the operation to conquer Singapore, when on 13 January 1942 the occupation of Kuala Lumpur convinced Yamashita that the Japanese would succeed in the Malaya campaign, he showed his original idea to Tsuji and, based on Yamashita's concept, Kunitake prepared the plan, being advised by Tsuji and Iketani, Section Chief of Operations, 25th Army. The plan of the Singapore campaign was actually prepared under Yamashita's supervision and according to his idea, except that the troops would cross the Johore Strait west of the causeway. As was touched upon before, the crossing area had been virtually decided as part of the plan of the Malaya campaign in Tokyo in Yamashita's absence.

Yamashita at once approved the plan of operations devised by Kunitake on 30 January. The following day, based on this plan, he ordered his assigned divisions to prepare to attack Singapore and, on 6 February, he gave them the attack order. Right after that, however, he himself changed the plan of operations.

First, he postponed its outset until the following day (at 24.00 on 8 February). The troops of the 25th Army drove south to Johore Bharu so fast that they were not fully supplied with ammunition and food, and the men, advancing night and day, had not recovered from their fatigue. Though they had been ordered to complete their preparation for attacking Singapore by noon on 7 February, few units met the timetable, causing Yamashita to postpone the attack by one day. As to this postponement, Tsuji greatly admired Yamashita's judgement and wrote that 'General Yamashita was considerate enough and was able to put himself in the position of those receiving his orders. He was a virtuous general by nature.'[44]

It was not clear whether Yamashita might postpone capturing Singapore on 11 February because his Army had arrived at Johore Bharu five days later than his original schedule. Before the outbreak of the war, Yamashita had intended to reach Johore Bahru in 50 days and to conquer Singapore on 11 February 1942, National Foundation Day.[45] He might prefer making sure of conquering the city rather than hastily attacking it, and his experience during the Malaya campaign might make him think that the British 'impregnable fortress' could be taken in three days.

Secondly, he changed the plan concerning the timing of the Imperial Guards Division's crossing of the Johore Strait. In the initial plan that Yamashita approved on 30 January, the 5th and the 18th Divisions would cross the strait in parallel on the front line and the Imperial Guards would follow them on the second echelon. But feeling that they would lose face, if they were held in reserve, the Imperial Guards appealed to Yamashita to cross the strait in the first echelon. Thus he changed the attack order he had already given and decided to use the Imperial Guards on the front line abreast of the other divisions.[46]

Yet, on 8 February, just before the outset of the operation of crossing the strait, the Imperial Guards demanded to be taken off the crossing because of their incomplete state of preparations. Yamashita refused, and strictly ordered the division to cross the strait the following night. The Imperial Guards obeyed and began their crossing at 23.00, 9 February. Soon afterwards the division demanded permission to suspend its crossing because one of its companies was allegedly annihilated by the British 'oil tactics'[47] and to advance instead in the rear of the 5th Division. Kunitake immediately went to evaluate the situation and did not find the losses so great as reported or the hindrance by oil on the water so serious as reported. Then he told the Imperial Guards to continue crossing the strait. This affair rendered Yamashita even more distrustful of Nishimura and his division.[48]

Even at the outset of the attack against Singapore, Yamashita's under-estimation of the enemy's strength had not been corrected. He still expected that the British would number 20,000 or 30,000 at most and that they would surrender if Bukit Timah fortress were occupied. He was rather concerned about the insufficient supplies of ammunition for his men.[49] At the beginning of the Singapore campaign, each gun could be allocated no more than 1,000 shells per week. To conceal the desperate shortage of ammunition from the British, Yamashita dared to continue to fire his artillery without hesitation from the opening of the campaign.

On 10 February, the Japanese occupied the British main defensive position at Bukit Panjang, and called on the British to surrender. The British coldly refused the appeal and only intensified their stubborn resistance. On 14 February, the Japanese front-line units had 100 to 200 shells left for each gun. Nevertheless, Yamashita firmly maintained his intent to continue firing, although some of his staff officers considered suspending the attack and waiting for supplies of ammunition.[50] At the same time, he ordered them to prepare for a general night attack at 20.00 on 15 February. He was about to realise the latter part of his original idea shown to Tsuji at Kuala Lumpur: if the British would not accept a call of capitulation, 'we would then openly carry out a battle of annihilation with all our military strength'. It was just then that the British agreed to a cease-fire. In retrospect, it can be concluded that Yamashita's conduct of war leading to the Japanese victory at Singapore served the purpose. But it is interesting to consider how things would have developed, if the British had continued their resistance for one or two more days.

Inside the 'joint operation'

In Japan, the Malaya/Singapore campaign has recently been recognised as a successful case of joint operation between the ground, sea and air forces.

When General Yamashita co-ordinated with the other forces, he told his counterparts all of his intentions regarding his plan of operations and about the support he hoped to get from them, but he never interfered in the details of how the other forces should act. By suggesting that he trusted his counterparts, he let them have confidence in himself. For example, Yamashita worked closely with Admiral Ozawa Jizaburo, Commander-in-Chief of the 1st South Expeditionary Fleet, responsible for naval operations in the South China Sea. Shortly before D-Day, the Army and the Navy Divisions at the Imperial General Headquarters could not agree about the plan of joint operations of the amphibious attack on the Malayan peninsula and left this issue to the two local commanders, Yamashita and Ozawa. When the two quickly concluded the local agreement in Saigon, Ozawa expressed his sentiment of sympathy with Yamashita by saying in a resolute tone: 'Land at Kota Bharu the way the Army hopes to. I take full responsibility for staking my fleet on supporting the operation.'[51]

As for communications between Yamashita's Command and the naval or air forces, there were few problems. The Navy dispatched Captain Nagai Taro as liaison staff officer to Yamashita's Command. Nagai was with Yamashita throughout the campaign, so they exchanged expectations and information on the spot without interference.[52] Good conditions for mutual understanding had been established.

From the standpoint of the 25th Army, the co-operation with air forces went relatively well from the support for the landing troops at the beginning of the campaign through to the bombing of Singapore in the final stage. The air forces responded well enough to the difficult demands of the 25th Army: for example, to prevent the bridges over the Perak River from being destroyed by the British (allegedly, Yamashita personally asked General Endo Saburo, Commander of the 3rd Air Brigade, to execute this operation), though it was not successful,[53] and to support Watanabe and Kunishi Task Forces sailing south along the western coast of the Malayan peninsula. During the Singapore campaign, the air forces provided much more air support than the 25th Army had expected.[54] Certainly, this is partly because co-operation of the ground forces to transport fuel and ammunition was essential for the air forces.

In comparison with the good relationships with the Navy and the air forces, those with the Southern Army[55] started to deteriorate shortly after the successful landings. On its second day, the Southern Army issued a citation to Takumi Task Force, which had succeeded in landing on Kota Bharu, but without any consultation with the 25th Army. On the issue of a citation, it was usual that divisions or armies should apply to the general armies for it or at least that the latter should confirm the will of the former. The 25th Army took offence at being ignored and at the wrong procedure followed by the Southern Army.

Secondly, the Southern Army demanded that the 25th Army release the 56th Division. That order was followed by the matters related to the cancellation of Operations Q and S. As mentioned before, the Southern Army insisted in executing these operations and would not easily agree to their cancellations despite the opposition of the 25th Army and the Navy. Besides, it expressed its regret over the 25th Army communicating directly with the Navy and Imperial General Headquarters and prohibited the 25th Army from doing so in future.[56] Captain Nagai's presence in the Command of the 25th Army worked well for mutual understanding and minimising difficulties for ground–sea co-operation. The Southern Army did not understand the effectiveness of such a relationship at all.

Finally, the Southern Army diverted air forces to the Sumatra campaign just before the start of the Singapore operation. This diversion was also done without any notice to the 25th Army. Tsuji wrote that 'General Yamashita looked as if he was displeased.'[57] The real intention of the Southern Army still needs to be examined, but it is likely that it was more interested in the Netherlands East Indies, the goal of the war, than Singapore, and that, to the Southern Army, the fall of the British fortress seemed to be just a matter of time.

To explain the bad relationship between the Southern Army and the 25th Army, Kunitake shed light on personnel affairs and insufficiency of contacts. Most of the staff officers of the 25th Army had been engaged so long in the Malaya/Singapore campaign – from the phase of operation planning – that they were familiar with details, while those of the Southern Army were not. So the former did not like the latter's interference and the latter did not like the former's attitude. Besides, the staff of the Southern Army visited the Command of the 25th Army only when they had something to do and they always left it soon after matters were settled. They never appeared when the 25th Army was engaged in hard fighting.[58]

Relationships between Yamashita's command and the Imperial Guards Division were not good, either. The division's way of fighting after their invasion of Singapore also irritated the Commander of the 25th Army. By letting the Imperial Guards attack the enemy in the flank, he intended to find a way out of difficulties for his main forces, that is, the 5th and the 18th Divisions. But the Imperial Guards moved too slowly to meet his expectations. Eventually, Yamashita lamented: 'What's taking the Imperial Guards Division so long?'[59]

The Imperial Guards Division had its say. Colonel Kunishi Kentaro, Commander of the 3rd Infantry Regiment, who actually received the orders in action, wrote that 'Yamashita's real intentions were not clear. It was hard for the men on the front line to understand them.' He thought that 'the main reason was the frequent changes of task (which meant objective of an attack) given to the Imperial Guards Division by the Army'.[60] In fact, after the units of the Infantry Corps of the Division occupied Bukit Mandai according to the initial operation plan, it was ordered by the corps headquarters to advance toward Changi on 11 February, and later on the same day was ordered by the division to advance in the direction of Nee Soon, and next day again was newly ordered by the division to advance to the south-east side of the Mac Ritchie Reservoir. Whenever the units received orders, they had to change their direction and to advance in the jungle throughout the night, fighting encounter engagements with the British.[61] What is more, it is not understandable that, after the end of the campaign, Yamashita seemed to think that 'the appropriate and effective employment of the Imperial Guards Division' was a reason for his victory.[62] This might have been so as not to make the division lose face.

Conclusion

Is it fortunate, or unfortunate, that Yamashita's decisions examined here probably did not have any great influence on the course of the campaign, except for his choice to drive rapidly toward the goal shortly after his landing on Singora? Is it possible to say that all his decisions were appropriate? In retrospect, since they did not prevent the campaign from being successful, none of them was irrevocably wrong. Therefore, it can be said that most of them were adequate for each situation and that his conduct of operations during the Malaya/ Singapore campaign probably deserves to be regarded as acceptable.

However, it should not be forgotten that Yamashita always underestimated the British military strength and it is undeniable that his impatience had some influence on his conduct of war. Moreover, it is certain that the warm-hearted general occasionally gave precedence to subjective human relations over objective analyses of the state of war. His regard for Mutaguchi worked well, while his consideration for Nishimura and the Imperial Guards Division often did harm, though things did not become crucial to the progress of the campaign.

Reflection on General Yamashita's conduct of war merely during the Malaya/ Singapore campaign does not make it possible to assert that he was an exceptional general in the Japanese Army. In practical terms, he only had to order and encourage his men to carry out operations in accordance with the plans devised by his staff officers. That is what Japanese Army commanders were commonly expected to do when campaigns went well.

7

BRITISH TACTICAL COMMAND AND LEADERSHIP IN THE BURMA CAMPAIGN, 1941–45

Graham Dunlop

Brigadier Willie Goschen and Lieutenant-Colonel Hedderwick now lie side by side in the Commonwealth War Cemetery at Kohima, just about where the front garden of the Deputy Commissioner's bungalow was before the battle. They died together in the monsoon mud on Kohima Ridge in May 1944. Goschen was a brigade commander in the 2nd British Division, which was then engaged in a hard battle to drive the Japanese from Kohima. Despite its superiority in mechanisation, artillery, armour and air power, the division was having a difficult time overcoming skilfully sited and stubbornly held Japanese defences. Goschen had led his 4th Infantry Brigade on a ten-day outflanking march through the mountains to attack the Japanese flank positions on Kohima Ridge at the foot of Pulebadze Mountain. After some early success, the brigade was held up by strong Japanese defences, and stalemate threatened.

Hedderwick was a 27-year-old Indian Army officer, commanding the 4th/1st Gurkhas in the 7th Indian Division, which had just been flown to Kohima after the recent battles in Arakan. There, the division had been surrounded, standing its ground for 17 days, supplied by air. During that battle, the Divisional Commander, Major-General Messervy, had been overrun in his own headquarters, and had to escape through the Japanese lines in order to rejoin his division.

Hedderwick's battalion had been sent forward to strengthen the attack on Kohima Ridge, passing through Goschen's brigade positions. Goschen and Lieutenant-Colonel Robert Scott, the Commanding Officer of the 2nd Battalion of the Royal Norfolk Regiment, were watching the proceedings from one of the Norfolks' forward trenches. Hedderwick's leading company was having a hard time and had lost all its officers, so Hedderwick decided to lead a new attack himself. As he moved forward, he was shot. Goschen's orderly ran forward and tried to drag him back to safety. He was shot. Goschen then ran forward and tried to pull Hedderwick in, and he was shot. Finally Scott, a giant of a man, charged forward and managed to pull Goschen in, but Goschen died in Scott's arms. Brigadier Theobald, who took over from Goschen, was killed three days later, in another attack on Kohima Ridge.[1]

Major Derek Horsford, also then 27 years old, took command of the Gurkhas after Hedderwick's death. In a later phase of the battle, he was ordered to clear the Naga Village on the other flank of the battlefield. Here was the final bastion of the Japanese defence of Kohima, which was very strongly held, and had resisted all attempts at frontal attack, inflicting heavy losses on the British. Horsford came up with a more subtle approach. By careful reconnaissance, good planning and excellent fieldcraft, his Gurkhas managed to outflank the Japanese position, using the Japanese technique of infiltrating by night through the jungle. Only when the Gurkhas were established there, and reinforced by tanks, which forced their way through to join them, did the Japanese position finally become untenable.[2]

I relate this episode because it will serve to remind us, during a clinical, academic observation of British tactical command in Burma, of just what it involved in that demanding theatre. It also points up a number of important aspects of tactical command and leadership, which characterised jungle warfare. First, we see how far forward commanders had to be: battalion, brigade and divisional commanders absolutely in the front line, indeed forward of it for some of the time. Hence, casualties among officers were high, especially as many were British commanding Indian or African soldiers, and were readily identifiable in the close-quarter fighting. Four brigade commanders became casualties at Kohima, of whom two were killed. Second, we might observe the youth of many of the tactical commanders. The environment and battlefield conditions made huge demands on them and they had to be fit enough to cope. Third, we see how the British command had to learn the tactics of the jungle, which did not come easily.

But this is only a glimpse of the subject. Now we must establish precisely what we are going to look at. The tactical level of command is concerned with the way weapons, equipment and people are used and sustained in battle, as well as maintaining the fighting spirit of the troops, in order to defeat the enemy. I define the tactical level of command in the British organisation during the Burma campaign as being that ranging from platoon at the bottom to corps at the top. Fourteenth Army and its predecessor, the Eastern Army, acted at the operational level. Successful command at the tactical level demands good tactical thinking and good management, as well as good leadership, and all three of these elements are inseparable and of equal importance. However charismatic, a leader will not inspire his subordinates if he manages them and their sustainment badly or directs them badly in battle. We cannot consider the quality of leadership in isolation from that of tactical thinking and management as well, so it is the way the British handled all these elements of command at the tactical level that I aim to examine.

I am not going to focus on any particular individuals, although I will name some as examples of a point from time to time. I want to take a broader look at how the British coped in general with the major command problems that faced them at the tactical level as the campaign developed. Without getting into a narrative of the campaign, I plan to undertake this examination chronologically.

At the end of 1941, when the first campaign opened, Burma was hopelessly ill prepared for war against a disciplined and experienced enemy. It was garrisoned by one division, which was established and trained primarily for internal security and ceremonial duties, and much of which was recruited locally. It had no armour and little artillery. Although an embryonic doctrine for jungle warfare was in existence, there had been little, if any, training in that environment, and certainly none for intensive operations. Operational responsibility for the Burma garrison had meandered between India Command in Delhi, Far East Command in Singapore and the combined American, British, Dutch and Australian Command, which formed in the Dutch East Indies at the start of the war. At about the time of the Japanese invasion of Burma, it reverted back to India Command. Meanwhile, administrative responsibility remained throughout in Delhi.[3] Hence, there was confusion and a lack of supervision of operational, tactical and administrative issues in Burma in the run-up to war. There was no accurate intelligence of Japanese capabilities or intentions, and there was a tendency before the war to underestimate the qualities of the Japanese soldier.

At the last minute, some battalion commanders undertook what little tactical and jungle training they could, despite lack of equipment, weapons and men. The 1st Battalion of the Gloucester Regiment, for example, was not involved in initial deployments to counter the Japanese attack in south Burma, having been held back to maintain public services in Rangoon. Many of the battalion's officers had been taken away to fill appointments in the Burma Army Head-quarters, which was not adequately staffed for war. However, at the last minute, the Commanding Officer, Lieutenant-Colonel Bagot, managed to organise some training in defence, making much use of expedients, such as string to simulate barbed wire. They had few machine guns or mortars, and no anti-tank guns.[4]

In January 1942, the Burma garrison was reinforced by the 17th Indian Division, which had just started training in India for mobile operations in the Middle East when the war against Japan started. Many of its junior officers, non-commissioned officers and experienced soldiers had already been taken away to strengthen units going to the Middle East, leaving it short of good junior leaders. Before being deployed to Burma, the division had been declared unfit for operations against a first-class enemy. Since then, two of its brigades had been detached to reinforce Malaya and had been replaced by others, with which it had not trained at all. It was all but destroyed as a fighting formation, being caught still forward of the river, when the Sittang bridge was demolished on 22 February 1942, and its Commander, Major-General John Smyth, VC, a sick man, was dismissed. The division was re-formed in March under a new Commander, Major General 'Punch' Cowan, the officer who, in his previous appointment, had declared it unfit for operations, and, under him, it fought with distinction throughout the remainder of the war.[5]

The brightest ray of hope in those early days was the arrival of the 7th Armoured Brigade, which had recent battle experience against the Italians and Germans in the Western Desert, and was a fully constituted, well-equipped and well-led combined arms formation. It was central to the successful withdrawal of the Burma Corps into India.

For largely political reasons the British had been ordered, against the better tactical judgement of divisional and brigade commanders, to defend too far forward, where communications were poor and their forces too dispersed in front of the natural lines of defence. For this reason they were already badly damaged by the time they fell back to the last good line of defence before Rangoon: the Sittang River. The British did not use up-to-date methods of mission command, so they lost the initiative through delays caused by a continual need for detailed orders downwards and for decisions to be referred upwards, as well as interference by higher commanders in the tactical decisions of their subordinates. In the early stages, for example, the 17th Division was receiving and issuing new operation instructions almost on a daily basis.[6] After the fall of Rangoon, the British were unsure of their mission; was it to deny upper Burma to the Japanese or to withdraw as much as possible intact to India?[7]

Unlike the Japanese, the British viewed the jungle as hostile and impenetrable. Thus tied to the few roads for movement and sustainment, and lacking any air cover after the first few weeks, the British never managed to counter the Japanese offensive tactic of outflanking enemy defences by infiltrating through the jungle and then cutting their line of communication. Every time this happened, the British withdrew, having to fight their way out through the road block, and often having to abandon heavy equipment. The tactic worked all the way down through Malaya and all the way up through Burma.[8]

Despite this failure of tactics, the first campaign did demonstrate a small number of important success stories in British tactical command. First was the speed with which the 17th Division was reconstituted after the disaster at the Sittang bridge. By early March 1942 it was back in the line under its new Commander, albeit at reduced strength, and it remained in action continuously until arriving back in India. Secondly, despite desertions by numbers of Burmese troops and failures, mainly among administrative, rear area and non-combatant personnel, the morale of British and Indian fighting troops appears to have been maintained on the whole.[9] Sir Reginald Dorman-Smith, the Governor of Burma, wrote in his official report after the evacuation of Burma: 'If guts and courage alone could have stemmed the Japanese invader, then Burma would still be ours today.'[10] Thirdly, the administrative staff had built up sufficient stocks in north Burma to sustain the garrison once Rangoon had been lost.

As a result of these things, as well as the tactical effectiveness of the 7th Armoured Brigade, the Burma Corps, formed from the 1st Burma and 17th Indian Divisions under Lieutenant-General Slim in March 1942, achieved that most difficult of operations: a withdrawal, in contact with the enemy, of over 1,400 kilometres. Despite heavy casualties and having had to abandon its equipment, it had maintained its cohesion as a fighting formation. That it managed this from a scratch beginning, with neither air cover nor the benefit of falling back on its main administrative base, reflects credit on the quality of tactical command at a time when Britain otherwise had little reason to celebrate its military prowess. Undoubtedly, it helped that Slim and his two subordinate division commanders, Major-Generals Cowan and Scott, were old Indian Army colleagues and friends, who understood and trusted each other implicitly.

A dismal postscript to the first Burma campaign was the very poor reception given by the Eastern Army of India to the Burma Corps as it reached Imphal. Woefully inadequate administrative arrangements and provision for accommodation had been made, and the arriving units were left to fend for themselves. This added considerably to the problems already faced by the leadership of the corps in maintaining the morale of their men after the long retreat and with the prospect of having to defend Imphal through the monsoon. It was a lamentable failure of management and basic human concern for the troops of the corps by the commanders and staff responsible.[11]

At the end of the first Burma campaign, in May 1942, it was clear that the British command had some major problems to solve at the tactical level before they could match the Japanese. First, they had to master the jungle so that they could live, move and fight in it. Officers and soldiers would have to be a great deal tougher and fitter. Concealment, control of fire and movement in battle, and disease prevention would require much greater discipline. Everyone would have to be able to navigate. Units would have to learn to operate away from roads, isolated and without mechanical transport, supplied by pack animals and air. Everyone, in both combat and administrative units, would have to be able to fight effectively, if only in self-defence. These things would require action at every level of command to develop the right training for individuals and units.

Second, in order to speed up command processes to win and retain the initiative in battle, where communications were poor, the British would have to master mission command techniques, using clearly understood common tactical doctrine and battle drills.

Third, the British had to develop a successful counter to Japanese infiltration and outflanking tactics, which they had failed to do throughout the first campaign. The British should not have been surprised by this sort of tactic, as there was nothing new about it in the history of warfare. Its devastating success against the British early in the Second World War was due largely to British reluctance to use the jungle and fear of being cut off while dependent upon roads for sustainment. In his report following the withdrawal into India, General Alexander, the Combined Burma Army Commander, wrote:

> The right method of defence was, I am convinced, to hold defended localities well stocked with reserves of supplies and ammunition, covering approaches and centres of communication, and to have behind those defended localities, hard hitting mobile forces available to counter attack the enemy should he attempt to surround the defence.[12]

These tactics would require sufficient troops to constitute the mobile reserves in depth, which the British could not afford during the first Burma campaign. In addition, apart from reducing their dependence on roads, the British would have to overcome their fear of being surrounded, and this was linked to the fourth vital factor: morale.

The maintenance of morale in South-East Asia was fraught with a number of problems, particular to that theatre, which were to remain a challenge for the command throughout the war, but which became apparent at this early stage. The environment was especially harsh, ranging from arid semi-desert to jungle-covered mountains with the highest rainfall in the world, as well as a number of virulent and dangerous diseases. The jungle, in particular, was strange and threatening to many of the troops, who, as we have seen, had not been trained for it at the start of the war. The Allied armies comprised soldiers of various different nationalities and cultures, who were unfamiliar to each other, having different diets, festivals, habits and conditions of service, but all of whom were mutually dependent and had to be welded into a cohesive fighting organisation. For the Allies, Burma was a subsidiary front, and held a low position in war priorities. Hence, equipment was late in coming, people at home were unaware of events in the theatre and the troops felt themselves to be forgotten. Growing socialist and anti-colonialist thinking in Britain and the nationalist movement in India threatened to sow seeds of doubt in the minds of troops about what they were fighting for. Soldiers were worried about conditions at home and the well-being of their families. On top of all this, the enemy was developing a reputation for ruthless invincibility.[13]

Despite the monsoon and the unhelpful reception given to units withdrawing from Burma, there was energetic command action on the IV Corps front at Imphal to address these issues. Major-General Cowan initiated urgent measures to get the 17th Division ready for immediate further action, even as his troops were sorting themselves out after their retreat. He commissioned a detailed report on the lessons learned from the first Burma campaign, the Cameron Report (so named after its author, Brigadier Cameron),[14] and directed intensive jungle, tactics and weapon training, as well as exercises to toughen junior officers and soldiers.[15] Cameron noted the same conclusions as Alexander on how to deal with the Japanese offensive tactics of outflanking and cutting British lines of communication, but there was little that could be done about them until forces were available and the doctrine, training, morale and supply arrangements were adequately developed. In addition to this intensive training, operations were mounted on the Chindwin River and in the Manipur Hills throughout the 1942 monsoon to maintain contact with the Japanese, gather intelligence, develop jungle tactics and build the confidence of the troops.[16]

XV Corps, based in Bengal and manning the Arakan front, was diverted during 1942 to internal security during the 'Quit India' movement, which interfered with training and created concerns over the reliability of Indian troops faced with this insurgency against British rule. Much of the small, pre-war, highly professional Indian Army had been deployed to the Middle East and Malaya, and it was in the middle of a rapid expansion programme. There were uncertainties about the political leanings of new recruits, and these concerns were aggravated by the formation of the Indian National Army in Malaya, which was recruited from Indian prisoners taken by the Japanese in the Malayan campaign to fight against British rule. British military leadership of the Indian Army had suffered a huge loss of credibility and face through defeats in Malaya, Burma and the Middle East, and most of those who joined the Indian National Army did so because they felt let down by their British commanders.

During this period, many Indian as well as British battalions had to be employed in support of the police, maintaining public order and protecting key points against attack by nationalist insurgents, and Indian soldiers faced great social and political pressures. In the event, however, the Indian Army proper remained largely aloof from politics and faithful to British command. This was attributed mainly to the strength of its regimental culture; the inherent loyalty of Indian officers and soldiers to their oath of allegiance; and a feeling among many Indian officers that independence would follow victory, at which time a properly functioning army would be essential to stability.[17]

Preparations for war, however, suffered from the diversion, and, despite very substantial local superiority, the British offensive on the Arakan coast over the winter of 1942–43 turned out to be an unmitigated failure caused by poor planning, training, tactics and leadership, leading to a collapse in morale. The plan had envisaged an amphibious descent on Akyab, but, when the landing craft were denied, it reverted to a straightforward frontal advance with tenuous, inadequate lines of communication. The formation and unit commanders involved did not have the experience of the first Burma campaign, and the lessons of that campaign were overlooked or ignored; jungle skill was still lacking and the British were still too dependent upon road sustainment. After years of garrison duty, many British officers were unfit for operational service.[18]

The British advance was ponderous, hampered by the need to build supply roads, allowing the Japanese ample time to discern British movements, reinforce and counter-attack. At Rathedaung and Foul Point the British found vital ground unoccupied, but failed to grasp the opportunity to seize it before the arrival of Japanese forces, and lost the initiative. Thereafter, British offensive tactics ignored the Japanese example of infiltration and outflanking, and were conducted frontally against strong defences without success and at unsustainable cost. After repeated attempts to overcome the Japanese defences, rapidly constructed at Rathedaung and Donbaik, had been repelled, there were reports of British and Indian units refusing to continue the attack. Despite the lessons from Malaya and Burma that armour could be used in the jungle, it was deployed just once, and then only in troop strength, in poor co-ordination with the infantry, and was knocked out quickly. The Japanese still managed to outmanoeuvre the British in the jungle and cut their lines of communication, inevitably forcing British withdrawal. British and Indian troops became fearful of the Japanese in the jungle, and there are reports of officers finding themselves deserted by their men.

The command arrangements for the operation were flawed. Major-General Lloyd's 14th Indian Division, the attacking formation, was launched under the direct command of the Eastern Army, commanded by General Irwin, who, in his previous appointment, commanding IV Corps, had been responsible for the inadequate reception of the Burma Corps at Imphal in May 1942. Slim's XV Corps headquarters, which would have been better equipped to handle the campaign and make a proper interface between army and divisional levels of command, was cut out of the loop. Consequently, the 14th Division took more and more brigades under command as they were deployed to strengthen the effort at the front, until the load became unmanageable for one divisional commander and staff. At that stage, Irwin went forward and took direct personal command of the division, which only made matters worse. Once it had become clear that the operation had failed completely, Irwin withdrew again and finally sent forward the 26th Indian Division, under Major-General Lomax, with Slim's XV Corps headquarters to take tactical command. They were just in time to salvage something of an orderly withdrawal. Even then, administrative command was retained by Eastern Army instead of being delegated to XV Corps along with tactical responsibility, a separation of functions that was unnecessary and could have proved fatal. The British retreated to their start point at Cox's Bazaar and were saved further embarrassment by the onset of the 1943 monsoon, but that withdrawal, in itself, required some deft tactical footwork and strong leadership by Slim and Lomax to avert a disaster.[19] The 14th Division reverted to a training role and never fought again. Irwin was dismissed.

In an appreciation of the situation late in the campaign, Slim reiterated the importance of defeating Japanese envelopment tactics by the surrounded unit standing and fighting, while mobile reserves dealt with the enemy encirclement, the tactics previously identified by Alexander and Cameron.[20] It was clear, however, that the resources, training and strength of leadership required were still lacking. In addition to the problems of jungle training, defence and morale that we identified earlier, there now emerged a new one: how to over-come Japanese defences, which were inevitably skilfully sited, strongly built and stubbornly held. Artillery and air bombardment had failed to suppress or destroy them, and frontal attack by infantry without armour had been repulsed with unsustainable cost in lives and morale. A solution would require the British to adopt the Japanese technique of infiltration, envelopment, cutting the line of communication and attacking from the rear. This, in turn, would demand much improved skill and confidence in the jungle, once again discarding the British dependence on roads. Where such techniques could not be used, the British would have to develop better combined arms tactics to co-ordinate the application of infantry, artillery, armour and air power, and to concentrate overwhelming firepower at the decisive point and time to break the Japanese defence. Until British commanders learned to use armour in the jungle, in sufficient strength and in intimate co-operation with the infantry, this would elude them.

Meanwhile, after the Arakan debacle, the morale of the British and Indians, and the confidence of their allies in the will or capability of the British to fight the Japanese reached an all-time low. There was a widespread mutual break-down in confidence between the troops and their commanders.[21] It represented a complete failure of command, and Churchill wrote:

> This campaign in Burma goes from bad to worse, and we are being completely outfought and out-manoeuvred by the Japanese. Luckily the small scale of the operations and the attraction of other events have prevented public opinion being directed upon this lamentable scene.[22]

One of those other events was the return of Brigadier Wingate's first Chindit expedition in April 1943. His 77th Brigade had not achieved any particular tangible operational success, save cutting the Japanese rail communications in the vicinity of Indaw and diverting a significant number of Japanese troops in pursuing his columns.[23] Even in the absence of any major battles, the 77th Brigade had suffered heavy casualties, mainly from disease, with only two-thirds of its number returning. However, the expedition was successfully portrayed as having achieved a moral victory, which eclipsed the failure of the Arakan offensive and set in train a gradual but relentless improvement in confidence among British and Indian troops that they could match, and eventually defeat, the Japanese in the jungle. The brigade was not composed of specially selected troops; indeed the British battalion of the King's Liverpool Regiment had been considered second-rate infantry, fit for garrison and security duties only. However, they had marched and skirmished their way 160 kilometres behind Japanese lines, supplied entirely by air. In many ways, for the British, the expedition was an experiment in jungle warfare independent of roads. It was characterised in large measure by the sort of adventurous, maverick leaders who are often suppressed in peacetime but who seem to emerge in every major war. They may not be war winners themselves, and there are many people who regard Wingate as a liability, who wasted good men on wild schemes. Nevertheless, such leaders often provide a pioneering tactical and moral example, which can help enormously in transforming a defeated, moribund army into one which wins battles. This was the true genius of the Chindit commanders.[24]

During the remainder of 1943, management and tactical thinking in the British command began to show marked improvement, despite the continuance of some of the problems that we have already identified as well as the emergence of new ones. A substantial boost to the quality and capacity of command at the strategic and operational levels came in October 1943, with the formation of the Allied South-East Asia Command and Fourteenth Army, backed by the training and administrative support of India Command. The energy they created percolated quickly down to the tactical level.[25] Training directives of formations within Fourteenth Army from about this time contain evidence of relentless emphasis on such matters as the toughening of troops, concealment, fire discipline, patrolling, methods of infiltration attack and combined arms tactics.[26] These memoranda and directives reveal a healthy interchange of ideas between field formations in the South-East Asia Command and the training staffs in India Command.

On the IV Corps front at Imphal, local jungle training and contact with the enemy at up to battalion level by the 17th, 20th and 23rd Indian Divisions continued to improve the skill and confidence of the troops.[27] Nevertheless, there was still some way to go in developing methods of overcoming Japanese defences. For example, a unit of the 17th Division was criticised heavily by a US observer for its unimaginative frontal tactics in an attack on Fort White in the Chin Hills in December 1943, albeit on very constrained mountainous ground with little scope for manoeuvre.[28] In other parts of that front, however, on more suitable ground, infiltration tactics were being used to better effect, to cut off Japanese defences and take them in the rear, avoiding costly frontal attack. One well-recorded example is the small battle at Kyaukchaw in the Kabaw Valley in February 1944.[29]

Although reports in the latter part of 1943 indicate that training and contact with the enemy were having a beneficial effect on morale, a new problem that became critical that year was the quality of officers and men being recruited into the British, colonial and Indian Armies. This was affecting all theatres of war owing to the increasing scarcity of suitable men. The Indian Army, for example, had expanded from some 190,000 at the beginning of the war to nearly 2,000,000 by mid-1943, and we have seen already some of the problems it had had to weather the previous year during the 'Quit India' campaign. On top of these, it had, by now, used up all the available manpower from the usual recruiting areas and classes, and was tapping regions with little or no recent military tradition. Many of the recruits were of poor condition. In July 1943, the Viceroy wrote in a telegram to the Secretary of State for India: 'It has become evident that a more extended training period is necessary due to a general lowering in the standard of recruits and inclusion of certain new classes inevitable with the expansion that has taken place.'[30] Special instructions had to be issued to all officers on the supervision of the diet and physical condition of new recruits.[31] In October 1943, the War Office Infantry Committee in London, considering the conditions of jungle warfare, noted that the demands of the more technical arms and services had resulted in the infantry being left with men unsuitable for their crucial role in jungle war.[32]

These, and the various other command problems we have seen, needed high-grade officers to address them, but there was, about this time, an increasing flow of complaints from commanders about the standard of new junior officers. In jungle warfare, where small units had to operate with a high degree of isolation, a great deal of trust and responsibility was thrust upon junior leaders, and commanding Indian or African troops, especially, placed further pressures on British officers. As well as winning their soldiers' trust through tactical competence and good management, they had to speak their soldiers' language, and they had to understand and involve themselves in the unfamiliar culture of their troops. Traditionally, this had demanded volunteers, who were carefully selected and then had to score well in training. On commissioning, they would remain in India or Africa for many years, developing a tightly knit professional and social cohesion.

The Indian Army, in particular, accrued extensive operational experience in the demanding classroom of the North-West Frontier of India. Consequently, at the beginning of the Second World War, it was highly professional, but had its own peculiar culture, which was not understood by all in the British Army. The rapid expansion of the Indian Army, coupled with inadequate selection and training early in the war, resulted in the appointment of many young officers, who were not volunteers, did not understand the culture and were of marginal quality. In particular, it was noted that many failed to take adequate interest in the welfare of their men, and increasing numbers of British officers joining Indian units were unable to speak the language of their soldiers. Considerable numbers of non-commissioned men were being sent for officer training, but there were complaints about the lack of preparation and suitability of many.[33] Ever greater numbers of Indians were being commissioned, and it was noted, not surprisingly, that Indian officers, having greater natural affinity with their men, got the best out of them more easily than did the newer British officers, but there were simply not enough of them.[34]

There were similar concerns about the quality of some of the British personnel being appointed to African units. A further problem here, the traditional practice of filling all command appointments with British officers or non-commissioned officers, was thought to sap the initiative of African soldiers, and this could have disastrous effects when their British commanders became casualties. This, again, was due largely to the rapid expansion of the African colonial armies, leading to a shortage of native non-commissioned officers.[35]

Measures to address these problems were beginning to show results by the beginning of 1944. By then, two training divisions, the 14th and 39th, were operating in India to put new recruits and units through intensive jungle training before they were accepted for active service.[36] New Indian Army officers' schools were established and increasing numbers of Indians were commissioned. Even if the quality and training of new British officers still left something to be desired, greater care was taken to integrate them into the Indian Army at unit level, as the following story of one Lieutenant Gadsden relates:

> Bangalore Garrison was not a friendly place for young officers. The senior Indian Army officers who ran the training establishment were absorbed in their own family circles, based on bungalow and club, and cold-shouldered the 'temporary gentlemen' passing through. On joining his battalion of the 14th Punjab Regiment, Gadsden found himself in a different world. First, he had to meet the Indian senior ranks. His Subedar was a Punjabi Mussulman, and from him he learned to respect the religious customs of his men . . . The Subedar encouraged the young British officers to mix socially with their men in a way unknown in the British Army. Gadsden drank tea and ate sweetmeats in the soldiers' lines several times a week. The young jawans responded, gathering round to sing songs and perform their folk dances at camp fires as the battalion trained in the Himalayan foothills . . . After a few months up in the hills the transformation was complete. As Gadsden writes: 'It took but a month or two for all of us to become dedicated Punjabis, and the officers who took the battalion to war the following year had built up the morale to a high point in the intervening training period.'[37]

Consequently, South-East Asia Command and India Command reports from early 1944 onwards indicate improvement and reducing concerns in these areas, and that the morale and confidence of the Fourteenth Army had reached a high level.[38]

By the summer of 1944, aided by the development of Allied air power and a growing numerical superiority of men and equipment, the British command had learned, and was putting into effect, the vital tactical lessons, which were identified previously. First, they had learned how to stand and fight, supplied by air, when cut off by the classic Japanese outflanking movement, allowing mobile reserves to destroy the attack and break the encirclement. Apart from the tactical and technical capabilities involved, this required a step change from the previous year in the attitude and confidence of the men thus surrounded, and it put considerable pressure on the leadership qualities of commanders until it could be demonstrated to work. It was achieved first by Messervy's 7th Indian Division, surrounded by the Sakurai Column of the Japanese 55th Division, during the battle of the Ngakyedauk Pass in Arakan, in February 1944. The possible necessity for air supply had been foreseen and prepared, so it worked well when the division was enveloped. The division also had the firepower and protected mobility of medium tanks within its perimeter.[39] It must have helped that Messervy himself had survived a similar experience fighting the Germans in the Western Desert, and he was undoubtedly a robust leader, who inspired confidence among his subordinates. As a result of his division's stand, the Japanese 55th Division was denied supplies, which it had planned to capture, and was successfully counter-attacked. Thus was achieved the first significant tactical victory over the Japanese by the British, and Slim regarded it as the turning point of the Burma campaign.[40] It must be said, however, that the local numerical advantage of five to one on the ground enjoyed by the British in this battle would have made British failure nothing short of a disgrace.[41]

The second such stand was that of the 1,500 men of the Kohima garrison, surrounded by the whole of the Japanese 31st Division, between 6 and 20 April 1944, during Operation U-go. This was made possible by air supply, artillery superiority, the developing counter-attack by XXXIII Corps and, most of all, the morale of the garrison. The only complete formed infantry unit in Kohima during the siege was the 4th Battalion of the Royal West Kent Regiment. This closely knit territorial battalion had recent battle experience, having been flown from the Arakan after the battle of the Ngakyedauk Pass, and was commanded by another particularly strong character, Lieutenant-Colonel John Laverty. Although there was friction between Laverty and Colonel Richards, the overall Commander of the Kohima garrison, there can be little doubt that Laverty's leadership, as well as the cohesion, moral strength and battle experience of his battalion, contributed overwhelmingly to the successful defence. The rest of the garrison was made up from the survivors of units already badly damaged by previous actions, as well as small *ad hoc* groups of men from administrative units, convalescents and reinforcement drafts, who would have been unable to form a cohesive defence by themselves. The West Kents had only arrived just in time, as the Japanese invested the town.[42]

Further south at this time, the whole of IV Corps at Imphal was cut off, but continued to fight, sustained entirely by air from 6 April to 22 June, when road communications were reopened by XXXIII Corps' advance from Kohima. Never before, or since, had such a large British formation achieved this. In terms of command, success was due in large measure to Admiral Mountbatten's efforts in securing sufficient aircraft to maintain the lift, but much credit must go also to the winning combination of some vigorous divisional commanders and the clear-thinking, if less charismatic, corps commander, Lieutenant General Scoones. These successes in defence provided a decisive morale boost to British and Indian troops, which enabled them to go on to the offensive imbued with the right confidence and fighting spirit.

Second, the British were finally getting the idea of using the Japanese offensive technique of infiltration to outflank the enemy and cut his line of communication, forcing him to counter-attack on less favourable terms than if he fought from his prepared defences. By 1944, with much improved confidence in the jungle and air supply following the Chindits' lead, this technique had been firmly embedded in British tactical doctrine, and was being taught, both in the training divisions and in the field army.[43] We have seen the effect achieved by the infiltration of Horsford's Gurkhas in the closing stages of the Kohima battle.[44] Meanwhile, during the concurrent battle on the Imphal plain, such outflanking tactics were being attempted with increasing regularity by both sides. One of the better-recorded British examples is that of the 48th Brigade's hook to cut the Tiddim road behind the Japanese 33rd Division fighting to gain access to the plain at Bishenpur. The block was kept in place for five days against repeated counter-attacks, inflicting much damage on the 33rd Division. However, it failed to precipitate a Japanese withdrawal, and this was indicative of the stubbornness of Japanese defence at the front, the defeat of which would often require application of the third lesson: overwhelming force applied through the use of combined arms tactics.[45]

At the time of the first Arakan campaign, over the winter of 1942–43, the British command, despite the advice of the Armoured Corps, ignored previous lessons and made pitiful use of tanks in 'penny packets', without proper co-ordination with the infantry, losing them to no avail. By 1944, fired by the Australian example from the New Guinea campaign, Slim, by then commanding Fourteenth Army, was determined that they should be used properly, and considerable experimentation and training in combined arms tactics had been undertaken. At Imphal, the 23rd Indian Division had focused on training with armour while the 17th and 20th Divisions were deployed forward in the Manipur and Chin Hills, and its role was to be the mobile reserve for the forthcoming Imphal battle. Medium tanks were deployed secretly to Arakan for the 1944 campaign, and two squadrons were within the 7th Division's perimeter during the battle of the Ngakyedauk Pass. They proved instrumental in enabling the infantry to counter-attack Japanese positions, and especially in destroying or suppressing strong Japanese defences by close-range direct fire.[46] At the same time, the co-ordination of air and ground operations was much improved by the integration of air and ground command at Army and Corps level. A staff officer from 11th Army Group was able to report thus after visiting the operations in Arakan: 'Fire support techniques have obviously been carefully studied and the application of combined support is being carried out smoothly, effectively and efficiently.'[47]

Later, when the relief of Kohima turned into a battle of attrition, after early attempts to outflank Japanese positions failed, it was only the combined weight of artillery, air power and tanks that enabled the British infantry eventually to overcome many of the Japanese defences.[48] The 2nd British Division was criticised for being too slow in clearing Kohima, and even some veterans of the division concede that it was ponderous, but, given the constraining nature of the ground and the strength of the defence, it is not surprising that the battle took so long. The 2nd Division had been equipped and trained for amphibious operations, and had been thrown into the Kohima battle to meet the crisis there. It was the only division in the Assam battles not to have seen recent action against the Japanese; indeed most of it had last been in action in France in 1940. Although Horsford's infiltration had a decisive effect in the final stages of the battle, there was little alternative to brute force in clearing many of the positions along Kohima Ridge, as the Japanese had found when they were attacking the town. Despite the overwhelming superiority of British firepower, casualties were heavy and the fighting conditions in the monsoon were appalling, and it is testament to the leadership of the division and the morale of the troops that they continued to fight so hard for so long. Nevertheless, at least one British battalion commander was relieved when his battalion faltered badly after he had refused to let them relax from full alert for days on end. Lieutenant-General Stopford, commanding XXXIII Corps, was under great pressure from Slim to get on, as, despite the airlift, the supply situation at Imphal was becoming increasingly critical. Relationships between Stopford and Major-General Grover, commanding the 2nd Division, became increasingly strained as the battle dragged on, especially while the 2nd Division was the only division in the corps. Once the 7th Division had arrived, Stopford had a full corps to command and he did not oversee Grover quite so closely, but their relationship continued to deteriorate.[49] Grover was dismissed after the battle, and that caused much disquiet among the men of the 2nd Division, who held him in high regard, and felt that the manner of his going slighted them after a hard-fought and vital action.

The crowning indicator that the British command had finally mastered both infiltration and combined arms tactics in the jungle came with the 5th Indian Division's advance back down the Tiddim road from Imphal, starting in July 1944.[50] Slim described it thus:

> Many of the Japanese positions were of great natural strength and all were stubbornly defended, but our troops ejected the enemy from each in turn. The method followed a pattern. As soon as the position was located, it was shelled and strafed from the air. While this preparation was going on, a wide outflanking movement would be launched through the hills to strike behind the enemy. Then, in co-ordination with this, a frontal attack with tank support would be launched.[51]

As this advance took place in the monsoon, the road line of communication was abandoned because it could not take the traffic. Unable to build airfields in the Chin Hills, the division was supplied entirely by airdrop and had to carry its casualties with it. A number of female nurses volunteered to go forward with the division to look after those casualties. Thus were the jungle tactics and methods, pioneered by the Japanese and Wingate's Chindits, being used by Fourteenth Army in concert with combined arms tactics. British tactical capability, and the command that led it, had come a very long way from the retreat of 1942.

Mission command, not used by the British in 1942 and 1943, was favoured by Slim, largely because of the environment and the wide dispersal of his forces.[52] In the events of early 1944, however, it failed on two occasions. The first was the timing of the decision to withdraw the forward divisions of IV Corps into the Imphal plain to fight the 15th Army there during the U-go offensive. Slim delegated the decision to Scoones, the corps commander. He, in turn, relied largely on the advice of his divisional commanders, who were naturally reluctant to give up ground for which they had fought, and were not in the fullest intelligence picture. Consequently, the decision was left too late, resulting in the 17th Division having to make its long withdrawal from Tiddim in contact with the enemy.[53] At about the same time there was a lack of clarity in intentions and the direction given for the defence of Dimapur and Kohima. This led to confusion over the positioning and tasks of reinforcements as they arrived. Consequently, Kohima was very nearly lost completely when the Japanese 31st Division arrived there on 6 April 1944.[54] Characteristically, Slim blamed himself for both these errors, but some responsibility must attach itself to his subordinates for not appreciating the situation clearly.

By the time of the Irrawaddy River battles of early 1945, mission command was clearly well understood by the corps and division commanders, as was demonstrated in the operations to cross that river and seize Mandalay and Meiktila. To illustrate the point, Fourteenth Army's operation instruction to IV Corps for the complex mechanised and airborne outflanking move to cross the Irrawaddy and seize Meiktila runs to just two sides of paper. The operation, clearly understood by subordinates, went off according to plan.[55]

By that late stage of the campaign, British command at the tactical level was clearly on top of the job that faced it. Whatever it lacked in tactical finesse was compensated for in the combat power and mobility that could now be deployed. The British were generally more at ease with the environment of the dry central plain of Burma, where they could use their superiority in mechanisation, armour and artillery to manoeuvre under total air supremacy. Hence, they were able to retain the initiative, outflank the Japanese defence, sever their lines of communication and then concentrate overwhelming combat power at the decisive point, almost at will. The art of tactical command by this stage was characterised less by the sort of cunning which had been required in the confines of the jungle, and more by the spatial awareness needed to co-ordinate a complex, multi-disciplinary, combined arms organisation, man-oeuvring at greater speed. What did not change, however, was the stubborn determination, skill and discipline of the Japanese soldier in defence. Thus, despite the machinery and scope for manoeuvre now available, the infantry and armoured troops were still faced with relentless, hard, close-quarter fighting in clearing Japanese positions, and that still demanded simple, uncompromising leadership.

Towards the end of the campaign, some cracks began to appear in the morale of British troops, caused mainly by war-weariness as well as concerns over affairs at home and repatriation. Large numbers of men had reached the end of their overseas tours of duty, which had been reduced from five years to four, and then, unexpectedly, to three years and four months. Hence, Fourteenth Army was faced with the immediate, unplanned departure of many more men than had been anticipated. As they were repatriated, they were replaced by inexperienced men, and the effectiveness of British units suffered. An increase in the number of complaints from soldiers about personnel management issues and the care shown by officers for the welfare of their men is evident from the morale reports of this period. This did not, however, appear to dent the confidence and fighting spirit of the troops as they sensed victory within their grasp.[56]

To summarise, therefore, British tactical command at the outset of the Burma campaign was hamstrung by the incompetence of previous preparations at the strategic and operational levels of command. Tactical commanders had to face the consequences, and they did their best with inadequate resources and training, but were overwhelmed and out-fought by the Japanese. The leadership of the Indian Army did well to maintain their troops' morale and loyalty during the 'Quit India' movement, but events in the Arakan over the winter of 1942–43 were nothing short of disgraceful, the blame for which lies at every level of British command involved. The tactical level, although under-resourced in the machinery of command and control, and interfered with from above, had more than adequate forces. It cannot avoid its share of responsibility for ignoring the tactical lessons of the first Burma campaign, and conducting its battles incompetently and without resolve. British combat effectiveness and morale reached a low point about this time, largely because of the Arakan debacle, but also because of recruiting difficulties and the quality of new officers and men. After the autumn of 1943, we see a transformation in resourcing and the strength of command, leading to a rapid improvement in tactical competence, administrative management and morale. This began with the formation of South-East Asia Command and Fourteenth Army, as well as the improved support and training from India, and it cascaded down through the chain of command. It enabled Fourteenth Army to defeat the invasion of India in 1944 and carried it forward all the way to Rangoon.

8

JAPANESE WAR LEADERSHIP IN THE BURMA THEATRE

The Imphal operation

Kenichi Arakawa

Introduction

From the end of 1941 to the summer in 1945, the battles in the Burma theatre occurred mainly at the beginning of each year. This is because fighting was virtually impossible during the monsoon season, lasting from May to October. In 1942, the Japanese conducted an offensive against Burma that culminated in its occupation. In 1943, the British Army attempted to take Akyab but failed. Nevertheless, when Brigadier-General Orde Wingate's brigade crossed the Chindwin River, in the first Wingate expedition, this operation increased British morale, despite the large amount of damage sustained and the inability to achieve any strategic objectives. In 1944, the Japanese Army crossed the Chindwin River from the opposite side to the west bank in what became known as the Imphal operation. Ultimately, this operation would undermine Japanese belief in the invincibility of the Imperial Army and would accelerate the collapse of the Japanese defensive system in Burma. In 1945, during the Irrawaddy crossing, Allied forces launched a counter-offensive into Burma.

General Sir William Slim remained in command of the British and Indian forces for most of the Burma campaign. Meanwhile, in March 1942, Orde Wingate arrived in Burma where he remained in command of the unconventional operation troops until his death in an accident in 1944. In the battles of 1942 and 1943, Slim was fully aware of the weakness of British and Indian forces, particularly of their low morale, and of the corresponding Japanese strengths, especially their agility. Some have argued that Wingate and his Chindits restored the morale of the British and Indian force. In any case, Slim trained the British and Indian forces, provided adequate rations and demonstrated their strength.

By contrast, no Japanese general remained in the Burma theatre throughout the campaign. Lieutenant-General Mutaguchi Renya was the only general stationed for a relatively long period in Burma. He joined the operation against Singapore as commander of the 18th Division. After the surrender of Singapore, he was assigned to the Burma theatre where his 18th Division fought well. Later Mutaguchi became the Commander of 15th Army, replacing Lieutenant-General Iida in March 1943. The strategic plan for the Imphal operation and its execution took place under the leadership of Lieutenant-General Mutaguchi. In September 1944, after the Imphal operation had been cancelled in the preceding July, Mutaguchi was recalled to Japan and his name put on the reserve list. In January 1945, he was recalled to serve as the president of the primary officer candidate school. He was serving in this capacity when the war ended in August 1945.

Most information about the Burma Theatre in the Second World War comes from the official Japanese and British histories of the war, written by their respective military establishments. Yet important questions remain unanswered.

Japanese leadership in the Burma theatre is considered in this chapter. Specifically, why did the Japanese Imperial Army not take offensive action against India in 1942, when the operation was feasible and could have had a significant impact on the tide of the Second World War? Yet why did the Imperial Army later pursue an equally risky operation against India despite serious problems and little expectation of any serious influence on the war? The Japanese Imperial Army sustained huge losses in this operation and its failure accelerated the collapse of the Japanese defensive system in Burma. In short, the Japanese war leadership failed to take advantage of timing.

According to the official British war histories, the British Army saw the strategic-geographical importance of Burma in the war against Japan in terms of two issues:[1] (1) Burma provided an eastern wall to defend India and particularly its industrial areas; (2) Burma provided an Allied supply route to China. Allied strategy was to tie down the maximum Japanese troops in the China theatre.

Conversely, for the Japanese Army, Burma was the western wall securing the resources of South-East Asia from an Allied attack. In addition, it constituted Chiang Kai-shek's only land supply route from the Allies. Once Japan occupied Burma, Chiang would be isolated and his defeat greatly facilitated. Therefore, at the beginning, the Japanese Army did not plan to advance into India, a British object of defence. An attack against India was not anticipated until after the Japanese Army had occupied all of Burma in July or August 1942.[2] The Imperial General Headquarters had ordered the Southern Army Staff to plan for an offensive as a means to protect Burma against an expected Allied offensive after the monsoon season. However, this operation, called Operation 21, was postponed indefinitely. Another event soon overtook the plan, namely the battle in Guadalcanal in the Pacific Ocean was the reason why this operation was postponed. Planners considered Guadalcanal to be a more decisive theatre than India.

In 1943, Wingate's Chindits suddenly crossed the Arakan Zibyu Mountains. Lieutenant-General Mutaguchi, Commander of the 15th Army and former Commander of the 18th Division, had thought that a large division could not cross the Arakan Zibyu Mountains. Wingate's success gave Mutaguchi a new conception for an operation against Imphal. In August 1943, the Imperial General Headquarters ordered preparations for the operation and, in January 1944, after many twists and turns, finally ordered its implementation. Today very few Japanese or British experts rate this plan highly.[3]

Japanese military strategy failed to capitalise on the momentum from an occupation of Burma to attack India because there was no plan to advance toward India in the first Burma campaign. The Imphal operation was carried out in accordance with Lieutenant-General Mutaguchi's original plan without revision, despite considerable opposition. This essay will analyse the Imphal operation from the viewpoint of military rationality, concentrating on the leadership of Lieutenant-General Mutaguchi. It attempts to outline the essential characteristics of the Japanese military leadership at each stage in the decision-making process including planning, execution and cancellation of the operation. The discussion will be subdivided into two time periods: (1) the planning stage up until the initial approval for the Imphal operation, and (2) the execution of the operation until its cancellation.[4]

The Imphal operation: planning, execution and cancellation

Phase 1: The planning

Prior to the Mutaguchi Plan: plans for a campaign against India and the impact of Wingate's Chindits

Operation 21, a plan to attack eastern India, surprised Lieutenant-General Iida, Commander of 15th Army. Its author was Lieutenant-Colonel Hayashi, a member of the Southern Army Staff. Iida considered the plan unrealistic and worked to prevent its implementation.[5] It then occurred to him that if he made the plan even more ambitious, many divisional commanders would think it impossible and would also oppose it, leading to its cancellation. Therefore, Iida actually ordered his staff to draft an even bolder operation plan than Hayashi's. He showed that plan to his divisional commanders and asked their opinion. Most of them thought it impossible to execute.

Mutaguchi's reaction at this moment was important. Under the Southern Army plan, Mutaguchi's 18th Division would lead the attack against Assam. In contrast, under the 15th Army's plan, the 18th Division would serve as a rear echelon. Mutaguchi informed Iida: 'I think there is only a small chance of conducting this operation, because there is little remaining time to make a logistical route and prepare a supply system.' So, the 15th Army in charge

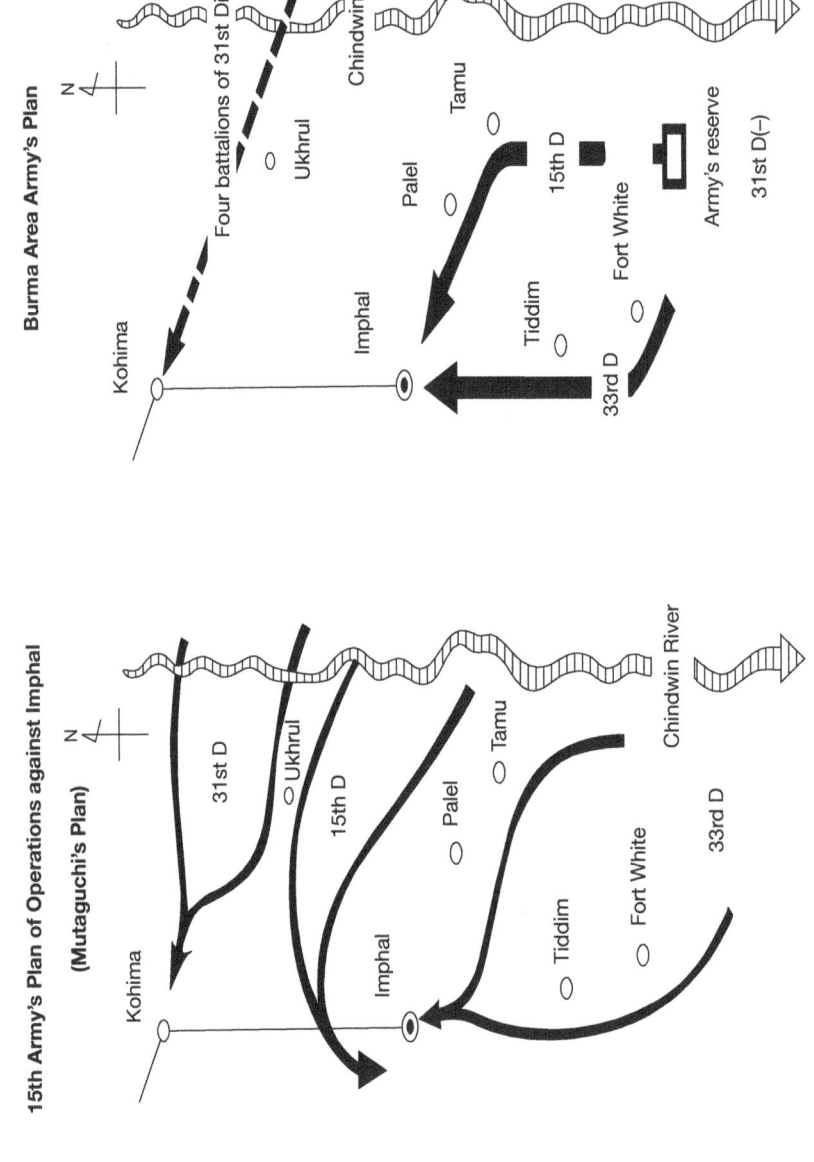

15th Army's Plan of Operations against Imphal

(Mutaguchi's Plan)

31st D
Ukhrul
15th D
Palel
Tamu
Imphal
Tiddim
Fort White
33rd D
Kohima
Chindwin River

Burma Area Army's Plan

Four battalions of 31st Division
Ukhrul
Chindwin River
Tamu
Palel
15th D
Fort White
Army's reserve
31st D(–)
Imphal
Tiddim
33rd D
Kohima

of the operation was not enthusiastic about it.[6] As the main battlefield had moved to the Pacific Ocean front, the operation was postponed indefinitely.[7]

At the beginning of 1943, Wingate's brigade succeeded in crossing the Arakan Mountains and the jungles below, shattering the commonly held assumption that it was impossible for a large force to cross such inhospitable terrain. Mutaguchi, then commander of 18th Division, received a report from a captain in Katha that a strong British force had crossed the Chindwin River and was advancing eastward. After this, Mutaguchi became fully engaged in a mopping-up operation against Wingate. (At that time, there was a large personnel change in the 15th Army with Lieutenant-General Mutaguchi's promotion to Commander of 15th Army.) According to Mutaguchi's reminiscences, Wingate's operation was traumatic:[8]

> The 15th Army acknowledged after this operation [the mopping-up operation against Wingate] that there had been a major error in geographic understanding: namely, the fact that we could pursue operations freely in this area, not only in the dry season but also in the rainy season. And furthermore, we could cross the Chindwin River easily by using materials [available] on the spot, such as dry-season rafts. We needed to change our prior thinking that, if we firmly protected the area along the mountain range, we could prevent an invasion by the enemy. This change in thinking altered the ordinary Japanese defensive strategy for central and northern Burma. At that time, I was really sure that, in case of a concentrated and repeated attack by additional echelons, the defensive line of northern Burma would verge on collapse. Therefore, I suffered for a few days until I was assured that no [other] echelons had followed Wingate.

From planning to approval: bypassing the opposition

Two events led to the acceptance of the Mutaguchi Plan. One was the above-mentioned unexpected attack by Wingate. The second was Mutaguchi's promotion in March 1943 from Commander of 18th Division to Commander of 15th Army. Originally, Mutaguchi had planned to make a pre-emptive attack on the enemy's counter-attack bases and then to pursue the original defensive role in his capacity as Commander of the 15th Army. This defensive objective grew and eventually evolved into a plan to reverse the tides of the Second World War by liberating India and compelling a British withdrawal. Furthermore, Mutaguchi was obsessed by his own personal ambitions: 'If I could decisively influence the Great East Asia War with an offensive operation against India, since I triggered it with the Marco Polo Bridge Incident, it would be very honourable for me.'[9] When the Marco Polo Bridge Incident had broken out, Mutaguchi had been the commander of an infantry regiment engaged in the fighting.

Mutaguchi requested permission to launch an offensive against Assam during a briefing delivered to Lieutenant-General Kawabe Masakazu, promoted to Commander-in-Chief of the Burma Area Army on 1 April 1943.[10] At that time Kawabe described this plan (hereafter referred to as the Mutaguchi Plan) as 'grandiose' and let it go in one ear and out the other. In other words, Kawabe, as Mutaguchi's senior officer, did not oppose the offensive against Assam. Afterward, Mutaguchi took every opportunity to promote his plan with his commanders in the Burma theatre and those in any way connected with its execution. Mutaguchi repeatedly requested the honour to put his plan into operation and deliver victory. There were only three commanders who could have stopped the Mutaguchi Plan: Lieutenant-General Kawabe, Commander-in-Chief of the Burma Area Army; General Terauchi, Commander-in-Chief of the Southern Army (and promoted to field marshal in June 1943); and General Sugiyama, the Chief of Staff of the General Headquarters (also promoted to field marshal in June 1943). Like Kawabe, all three generals were thinking of letting Mutaguchi have his way and therefore none worked to oppose the plan.

If there had been any opponents to the Mutaguchi Plan, it would have been from the staffs of the higher headquarters, because Mutaguchi did not brook opposing opinions from his own subordinates. Colonel Katakura, senior staff officer to the Chief of Staff of the Burma Area Army; Major-General Inada, Vice-Chief of Staff to General Terauchi at the Southern Army Headquarters; and Colonel Sanada, the chief of the 2nd Section of the 1st Bureau of Operations at the Army General Headquarters, might have opposed the Mutaguchi Plan.

Although Major-General Inada seemed to have been its key opponent, the only one to oppose the Mutaguchi Plan itself was Obata, Chief of Staff of the 15th Army. Katakura, Inada and Sanada opposed the specific offensive strategy, not the concept of an offensive operation in order to defend Burma.

The question then becomes: why was the Mutaguchi Plan never revised? Why was the issue of a non-total offensive by the enemy never considered? In reality, it was not necessary for the 15th Army to attack India because the main objective of the operation was to protect Burma.[11]

At the end of April 1943, the newly assigned Obata and other staff submitted a report recommending the cancellation of the Bu-go operation to set up a defensive line in Burma in the direction of the west bank of the Chindwin River. In response, for the first time Mutaguchi presented his plan to his general staff. Mutaguchi had ordered his staff to study and make plans to implement the Bu-go operation, claiming he intended to execute it just before the rainy season in preparation for an attack on India. Mutaguchi accused his staff of passivity, arguing:[12]

> The overall situation of the war is a stalemate. If there is any possibility of breaking out of this situation, it would be in the Burma theatre. Therefore, we should not be passive. On this occasion we should take

the offensive and make Imphal surrender. And furthermore, if possible,
I would like to lead my force to Assam. From now on, we should cease
making defensive plans and shift to offensive ones.

(In this speech Mutaguchi quoted the words of Major-General Ayabe, chief
of the 1st Bureau (Operations) at Army Staff General Headquarters (GHQ), to
imply that Ayabe had agreed to Mutaguchi's plan and had mentioned the
possibility of the addition of two divisions.)

Obata, however, tried to persuade Mutaguchi to abandon plans for an
offensive against India. Obata asked the Commander of the 18th Division to
persuade Mutaguchi to abandon his plan. Then Mutaguchi criticised Obata's
failure to support him. Thereafter, Obata was reassigned. Obata had been
assigned as Chief of Staff of the 15th Army only two months previously.
Meanwhile, there was a meeting of the commanders of the 31st and 33rd
Divisions, in which Mutaguchi outlined his plan against India, but the relevant
commanders opposed the idea.[13]

A war game was held at the Headquarters (HQ) of the Burma Area Army
in Rangoon at the end of June 1943. It was intended to create plans to protect
Burma. The 15th Army brought the Mutaguchi Plan to the war game, which
was conducted according to this plan.

Afterward, Chief of Staff of the Burma Area Army Naka evaluated the
results, saying, 'There were a lot of problems with the Mutaguchi Plan.' He
presented a more deliberate and secure Burma Area Army plan. Major-General
Inada backed Naka's evaluation. Inada's evaluation of the Mutaguchi Plan
seems reasonable. He commented:

> The Mutaguchi Plan is full of fond hopes. He counts his chickens
> before they are hatched. The Mutaguchi Plan goes like this: the Army
> crosses the Chindwin River and goes over the Arakan Mountains,
> where there is no road, with as much ammunition and food as it
> can carry. And then when rations run out, food and ammunition are
> to be gained in Imphal and supplied to the divisions, using as much
> transportation as necessary captured there. Such an operation might
> have been possible in the spring of 1942. But now that the enemy is
> preparing for a counter-attack, this plan is senseless. The plan presented
> by the Burma Area Army is feasible. In any case, the Mutaguchi
> Plan lacks flexibility. At this time, the operation is a means to defend
> Burma, not to make an all-out attack against India. Mutaguchi doesn't
> make this kind of acknowledgement.[14]

Inada decided not to let an unrevised Mutaguchi Plan (the Imphal operation)
be pursued. He explained his opinion about the Imphal operation and, when
he went to Imperial General Headquarters, he got tentative approval supporting his objections. Surprisingly, in August 1943, the Imperial General

Headquarters (IGHQ) still ordered preparations for this operation. At the time, Inada had just returned from his trip to IGHQ.[15] Under this IGHQ directive, the chain of command put the Southern Army in the position of directing the U-go operation (Imphal operation) and the Burma Area Army in a subordinate position. The Burma Area Army, in turn, commanded the 15th Army. However, the directive from the Burma Area Army was not the clearly revised version of the Mutaguchi Plan.

It is important to note that, in the preparatory directive, the Southern Army took the position that there would be no general enemy offensive. (Actually the Allies were about to change their plans from an all-out attack to an ambush and strike-back strategy.) But this possibility was not considered inside the Burma Area Army or the 15th Army. The thinking of the Southern Army did not percolate down to the 15th Army.

The war game in compliance with the orders to the Burma Area Army to prepare the U-go operation took place at a divisional commander meeting at the 15th Army HQ at the end of August 1943. Mutaguchi did not change his plan in this meeting. Naka acquiesced to the unrevised plan. Noticing Naka's implicit approval, Katakura pointed his finger: 'Acquiescence means official approval. No one can revise it once it gets approved.'[16]

The Southern Army held a Chiefs of Staff meeting in Singapore on 11 and 12 September 1943. The Chief of Staff of the 15th Army explained the U-go operation plan, which was in fact Mutaguchi's original plan. At the meeting, Vice-Chief of Staff of the Southern Army Inada requested revisions. Chief of Staff of the Burma Area Army Naka, however, supported the Mutaguchi Plan. Katakura argued: 'The attitude of the 15th Army indicates a disobedience to orders despite several recommendations from both the Southern Army and the Burma Area Army.' Kawabe refused to accept Colonel Katakura's suggestion, considering it interference in the 15th Army's operational plans.[17]

Inada was suddenly transferred on 15 October 1943 and replaced by Major-General Ayabe, chief of the 1st Bureau of the Army Staff GHQ.[18] In this way military rationality was undermined. At this stage, the Imphal operation (Mutaguchi Plan) was adopted virtually without revision.

On 15 December 1943, the 15th Army HQ conducted more war studies on the Imphal operation. The Chiefs of Staff of the Burma Area Army were the main participants in this study. The division commanders and key persons who would be in charge of the eventual operation were not invited. The purpose of this war game was not to evaluate the feasibility of the Mutaguchi Plan but to secure its approval. Ayabe (promoted to lieutenant-general in October 1943) attended this war game, approved the Imphal operation and, after many twists and turns, finally made a report about the plan to General Terauchi, Commander-in-Chief of the Southern Army. After gaining his approval, Ayabe went directly to Tokyo for the approval of the IGHQ. They said General Sugiyama persuaded Major-General Sanada, the chief of the 1st Bureau at the Army HQ, to approve the operation. Sanada's opposition was considered to be weak because the person seeking its approval was Lieutenant-General Ayabe, who was four years his senior as well as his immediate predecessor.

On 4 January 1944, Imperial General Headquarters approved the request and, on 7 January 1944, it issued the IGHQ Army Directive No. 1776 (approval of the Imphal operation):

> For the defence of Burma, the Commander-in-Chief of the Southern Army shall destroy the enemy on that front at the appropriate juncture and secure a strategic zone in north-east India in the area of Imphal.[19]

Thus the Mutaguchi Plan was approved for execution in its original form without any revision.

1943: Change in the position of the Burma theatre in British strategy, the tides of war, and political factors

In January 1944, when the Mutaguchi Plan was approved with no revisions after a year-long consideration period, Mutaguchi became more ambitious and optimistic than ever. After the British withdrawal from Burma in 1942, they had planned an all-out counter-offensive against Burma. Yet the British postponed plans for a major attack at the end of 1943. This represented a change in British strategy. In the end, the British remained on the defensive and tried to restrain the Imperial Japanese Army with the minimum military force.[20]

Prime Minister Winston Churchill issued a directive to Admiral Lord Louis Mountbatten in October 1943:

> Your prime duty is to engage the Japanese as closely and continuously as possible in order by attrition to consume and wear down the enemy's forces, especially his air forces, thus making our superiority tell and forcing the enemy to divert his forces from the Pacific theatre.[21]

In short, the Burma theatre was to be a supportive operation. The British hoped to use it to slow the advance of the Japanese Army and to draw Japanese forces away from the Pacific Ocean front. In November 1943, Admiral Mountbatten's plan was to restrain the Japanese Army in the Burma theatre with minimum force. Thus, the more enthusiastically the Japanese Army threw forces into the Burma theatre, the more advantageous it was for the British forces, who desired precisely this outcome.

By February 1944, the British caught the first signs of the impending Japanese Imphal operation. Under Lieutenant-General Sir William Slim's final operational plan, the British would be on the lookout for any movements by the Japanese 15th Army and destroy it around Imphal.[22]

The worsening situation in the Pacific Ocean theatre also contributed to the Japanese Army's decision to take the offensive in Burma. In 1943 reversals in the Pacific Ocean theatre had compelled the Japanese Army to take one defensive action after another. In the IGHQ, there was a growing sense of urgency to regain the initiative by taking offensive action somewhere. In October 1943, Colonel Hattori, the former secretary to the Minister of the Army, reassumed the post of chief of the 2nd Section of the 1st Bureau at Army GHQ, replacing Sanada, who was promoted to chief of the 1st Bureau. Hattori would plan the Ichi-go operation in order to break through from north to south China in 1944. Undoubtedly Hattori did not oppose the Imphal operation because its success would facilitate his own Ichi-go operation. Finally, the activities of the anti-British Indian Nationalist, Subhas Chandra Bose, also fed into plans for a Japanese offensive in Burma.[23] In July 1943, Bose called on General Kawabe. Bose's character fascinated Kawabe who assured his guest that Japan intended to launch a major offensive operation against India in the Great East Asian War.[24] Bose also met Prime Minister Tojo Hideki. In that meeting, Japan officially recognised the Free India Provisional government established by Bose in Singapore on 1 October 1943.

Thus, the worsening tides of the war and various political factors combined to propel the Imphal operation forward.

Phase 2: The Imphal operation from execution to cancellation

Lieutenant-General Mutaguchi's execution of the operation

Mutaguchi's flawed execution of the Imphal operation contributed to raising its costs and helped precipitate an early collapse of the Japanese defensive system in Burma. In particular, he delayed the cancellation of the operation for two months after he himself had lost any hope of its successful execution.

A Canadian scholar has criticised him for being 'incapable'.[25] Yet Mutaguchi had passed the extremely difficult entrance examination to the Army Staff College just after he was promoted to lieutenant. In his days as a junior officer, there are no doubts he was a first-class student. For 18 years, after he graduated from the Army Staff College, he had served at the GHQ and the Ministry of the Army. Before he was transferred to command an infantry regiment in China, he had spent less than two years as commander of an infantry battalion of the Imperial Guards Division. He seems to have been a typical military bureaucrat.

It is unclear whether or not his bureaucratic career path had weakened his sense of understanding of the battlefield and his operational leadership skills. There are indications of his limitations in these areas. For example, around 5 March 1944, just before the start of the Imphal operation, Mutaguchi sent a private letter to Tojo requesting him to provide additional anti-aircraft artillery to defend against airborne troops. At that time, Mutaguchi was already under attack by airborne troops and it was too late to make such a request.[26] This is just one indication of his unrealistic decision making.

Mutaguchi made repeated logistical miscalculations, which help account for his inability to see the weakness of his plan and flaws in its execution. There are also questions about his leadership ability. At the end of April, the 15th Army was to attack with the support of the Burma Area Army. Mutaguchi insisted on sticking to his plan to attack at Bishenpur, where the enemy had not spent much time making a fort and was rather unprepared. To reach Bishenpur, however, the 15th Army had to move tanks and heavy artillery from over 50 kilometres away at Palel without control of the air. Mutaguchi's plan failed to consider the time this would take, the damage that would be incurred or the labour entailed. Nor did he consider the risks resulting from the attack in the event of its failure: if defeated at Palel, the enemy would be able to threaten the rear of the main Japanese force. This would jeopardise the entire Imphal operation. Mutaguchi's leadership had numerous problems on the operational level.

THE LOCATION OF MUTAGUCHI'S HEADQUARTERS

From 8 March to 20 April 1944, Mutaguchi commanded his forces from Maymyo, 300 kilometres from the front line. Thus, he was far removed from the theatre during the crucial period when success or failure would be determined.[27] Until his HQ was moved to Indainggyi on 20 April, his orders often lacked comprehension of the battle-front situation. For example, the Right Assault Force of the 33rd Division that fought at the Tamu Palel front was put under the direct control of General Mutaguchi, Commander of the 15th Army, and became Yamamoto's Flank Column. At that time, the Left Flank Column of the 15th Division was transferred to Yamamoto's Flank Column. Once this command structure had been established and the orders issued, it was impossible to change during the heat of battle. In battle, such changes would have been far too risky and too costly. Then only two regiments (that is six battalions) remained in the 15th Division. After one battalion had been shifted from the 15th Division, this left only five battalions for the 15th Division. The 15th Division was supposed to be the main division in charge of the attack, yet it had the same force size as Yamamoto's Flank Column.

On 26 March 1944, Yamamoto's Flank Column, operating as a column under the direct control of the 15th Army, occupied Moreh. This did not represent defeat of the enemy forces since the latter had withdrawn according to plans without fighting. Reportedly, Yamamoto described his gain as a 'defeat'. Yet according to the records of the Imphal operation written mainly by Mutaguchi and his staff at the 15th Army HQ, Yamamoto's occupation of Moreh was a breakthrough.[28] Although Yamamoto's Flank Column was under the direct control of General Mutaguchi, he still did not understand the operational significance of its victory. This, in turn, demonstrates his lack of understanding of the battle situation.

THE ORDER TO RETREAT BY LIEUTENANT-GENERAL YANAGIDA, COMMANDER OF 33RD DIVISION

On 14 March 1944, the 33rd Division initially succeeded in cutting off the 17th Indian Division's retreat by taking advantage of the delayed withdrawal of this division. However, the force strength had been decreased for the 1st and 2nd Battalions of the 215th Infantry Regiment of the 33rd Division, the regiment in charge of cutting off the 17th Indian Division's retreat near Singgel. This was because the 215th Infantry Regiment was wedged between the 17th Indian Division and an additional British relief column from Imphal.

On 25 March, Colonel Sasahara, the Commander of the 215th Regiment, sent a telegram to Lieutenant-General Yanagida, saying: 'We are devoted to our duty, seeking an honourable death in a *banzai* charge.' At the same time, Colonel Sasahara ordered his two battalions to retreat on his own authority. Meanwhile Yanagida, the Commander of the 33rd Division, ordered Colonel Sasahara to retreat because he misinterpreted Sasahara's phrase, 'seeking an honorable death in a *banzai* charge', for 'they all died in a *banzai* charge'. Actually the 215th Regiment in charge of cutting the path for a retreat had already been exhausted in battle and would have been crushed by British and Indian forces in any case. This meant there was a shortage of Japanese infantry troops in Singgel to cut off the 17th Indian Division's retreat.

The fault lay in Mutaguchi's troop deployments in his plans for the Imphal operation. According to the Burma Area Army plan, the 33rd Division might have been able to send reinforcements to Tongzang-Singgel, because the 15th Division would be approaching Imphal through Tamu-Palel. Under this plan, the 31st Division would be kept at full strength as a reserve. However, the Mutaguchi Plan had split the 33rd Division, with one part going to Tongzang-Singgel while the second part went to Tamu-Palel. This meant that there was only a small infantry force deployed at the front and no infantry left in reserve. Under the Mutaguchi Plan, the 33rd Division could not send a relief column to the regiment in charge of cutting off the enemy retreat. This, in turn, allowed the enemy to break through and retreat intact. The evacuation of the 17th Indian Division was successful not because of Yanagida's order to retreat, but because of Mutaguchi's flawed troop deployments and the misinterpretation of Sasahara's telegram.[29]

THE BREAKTHROUGH TOWARD DIMAPUR

On 6 April 1944, reports reached the 15th Army HQ of the successful occupation of Kohima by the Left Flank Column of the 31st Division. Mutaguchi ordered the Commander of the 31st Division 'to continue to chase the enemy to Dimapur'. The Burma Area Army, however, immediately telegrammed an order 'to halt the operation by the 15th Army to pursue the enemy'.[30]

After the war, Mutaguchi claimed in the official document that, 'if the 31st Division had pursued the enemy to Dimapur, not stopping in Kohima, every

British attempt to relieve Imphal would have been crushed and eventually the British military system would have collapsed'. He based this claim on a 1962 letter by Lieutenant-Colonel Barker, who had served on the staff of the British 14th Army.[31] Mutaguchi was delighted to read Barker's letter, responding 'This letter is the voice of God.' 'I was not wrong. If I had done everything I had wanted to do, I would have won the operation.' He indicated his resentment of Lieutenant-General Kawabe, who had prevented the 31st Division's pursuit of the enemy toward Dimapur. In the same document, Mutaguchi wrote that Major-General Miyazaki, Commander of Left Flank Column of the 31st Division, had also admitted the wisdom of Mutaguchi's thrust toward Dimapur.

Was such a thrust toward Dimapur by the 31st Division really possible? Pursuit of the enemy toward Dimapur would have been very difficult to execute. The British and Japanese official histories of the war provide a compelling analysis of the Japanese and British advance to Kohima heights during the period from 4 April to mid-April 1944.

Arthur Swinson, the author of *Kohima*, compared the orders of Mutaguchi with those of Kawabe, concluding that 'Mutaguchi was right: there could be no doubt whatsoever' at the time.[32]

Yet this conclusion overestimates the capabilities of the Imperial Japanese Army. Circumstances at that time would have put the 138th Infantry Regiment at the forefront, capable of advancing. On 6 April, the 3rd Battalion of this regiment advanced to a hill 15 kilometres west from Kohima to disrupt the road between Dimapur and Kohima. They fought the British troops deployed there. Not only did their night attack fail, but their battalion commander also died. From then on, both sides remained locked in conflict there.[33]

Victory would have been possible only if the Japanese Army had been able to supply itself adequately and to provide the necessary artillery fire, both of which the British forces were able to do for themselves. The advance of the mountain artillery regiment of the 31st Division was scheduled for around 20 April 1944.[34] Meanwhile, the supplies carried in by the 31st Division were already running short, while supplies from the 15th Army were not expected and those seized from the enemy had been lost by a combination of gunfire and air raids. British reinforcements from Dimapur were pouring into the region. In 1944 this force was not the same British force that had rapidly fled when Mutaguchi had surrounded British forces to take Singapore.

Even if Mutaguchi's plans seemed reasonable in theory, in practice they lacked the necessary logistical support for success. His line on the map could be drawn in no time. It did not take into account the reality that armies are composed of humans who must be able to carry sufficient quantities of heavy guns, bullets and food to execute an actual operation.

The cancellation: reassignment of three divisional commanders and
delays in calling off the operation

REASSIGNMENT OF THREE DIVISIONAL COMMANDERS

Extraordinary things happened in the Imphal operation. All three divisional
commanders of the 15th Army were reassigned. Although the cases were
not identical, the three shared a common opposition to the Mutaguchi Plan
and Mutaguchi knew it. He ignored their objections and made no attempt
to persuade them patiently of the logic behind his reasoning; rather he imposed
his plan unilaterally. After 1943, the commanders of the 31st and 32nd
Divisions did not attend the meetings or war games related to the operation,
so they never heard Mutaguchi's explanations for the Imphal operation. It is
unclear whether the three commanders refused to participate or whether they
were deliberately excluded. In the normal course of events, Mutaguchi should
have ordered them to attend.

The three divisional commanders were reassigned after the trend of this
operation had been decided. First, in early May, the Commander of the 33rd
Division, Yanagida, was faulted for leadership problems and reassigned. Second,
the Commander of the 31st Division requested reassignment elsewhere. Third,
the Commander of the 15th Division succumbed to disease and was removed
from the operation. Mutaguchi may not have known about the illness and the
worsening health of the Commander of the 15th Division, Lieutenant-General
Yamauchi. Only Chief of Staff of the 15th Division Okada and the doctor
of the division would have known about the illness.[35] Therefore, there was a
possibility that Mutaguchi did not know about the illness when he reassigned
Lieutenant-General Yamauchi. The real reason behind Yamauchi's reassignment might be Mutaguchi's low opinion of the lieutenant-general's leadership
ability.

Thus, the 15th Army rushed into the Imphal operation without close ties
among its divisional commanders. Flaws in assessment, imperfect forces, shifts
in command and especially Mutaguchi's own flawed leadership greatly
increased the number of casualties.

DELAYING THE CANCELLATION

More than any other factor, delaying the order to halt the operation accounts
for the large losses and needless damage incurred. There were two opportunities
to halt the operation before the final halt order was issued on 3 July 1944.
One was at the end of April when the provisions and ammunition carried by
the divisions were running out. On 20 April, for the first time, Mutaguchi
went to the front to observe the actual situation. As soon as he arrived at
his command post at Indainggyi near the front line, he lost confidence there
because reality did not accord with his own prior beliefs. In a document of

a meeting that April between Mutaguchi and Ushiro, a staff officer of the Burma Area Army, Mutaguchi exclaimed: 'At the very last moment, we lack the necessary military strength. I have profound regrets.' Ushiro had come to Indainggyi to assess the situation. In addition, Mutaguchi wrote on his name card given to Lieutenant-General Naka, Chief of Staff of the Burma Area Army, 'Thinking of far distant Tokyo, I am very ashamed of myself.' At this time Mutaguchi admitted the failure of the operation.[36]

Mutaguchi's second chance to call off the operation came during a meeting with Kawabe on 6 June 1944. On 25 May 1944, about ten days prior to the meeting, Mutaguchi had received a telegram from Sato, the commander of 31st Division, saying: 'We are going to withdraw from Kohima to a place where we can get supplies.' Mutaguchi was surprised by this and had replied by telegram: 'It's hard to understand.' But in another telegram he did not order Sato to remain; rather, he implicitly gave approval to Sato's withdrawal on 2 June 1944.[37]

The meeting between Kawabe, Commander of the Burma Area Army, and Mutaguchi, Commander of the 15th Army, took place under these trying circumstances. The meeting itself was strange. According to Mutaguchi, '[The thought] "The time has come to give up the operation as soon as possible" got as far as my throat, but I could not force it out in words. But I wanted him to get it from my expression.' According to Kawabe, 'From his face I got the impression that he had something to tell me but he did not say anything. But I left him without asking him anything directly.'[38]

By postponing decision making, they both squandered the opportunity to save many soldiers' lives.

LOST LIVES FROM DELAYING THE CANCELLATION

At the time of the Mutaguchi–Kawabe meeting, troops in the front line continued their attack. At dawn on 7 June, the 1st Battalion of the 215th Regiment of 33rd Division started its second attack. The first unit engaged the enemy in hand-to-hand combat. Lieutenant Nagai Tatsuo, the Commander of the 4th Company; Second Lieutenant Abe Toshio, Deputy Commander of the 2nd Company; and Second Lieutenant Fujisawa Daijiro, Deputy Commander of 3rd Company, all died heroically in the concentrated enemy gunfire.[39]

One of these company leaders, Second Lieutenant Abe had just returned, a month after being ordered on 28 March to lead a volunteer raid to blow up a bridge with 28 soldiers. He had returned without losing any of his soldiers on the mission, but he would die needlessly in Burma.[40]

According to my research, a total of 13,376 officers and soldiers of the 15th and the 33rd Divisions died in the Imphal operation. More than 7,500 of them died after June 1944.[41]

Eventually the irrevocable fact that attacks by British and Indian forces had opened the Kohima–Imphal road resulted in the cancellation of the Imphal operation. The Imphal operation had failed to secure this vital objective. This time even Mutaguchi and Kawabe reported the implications to senior staff. The Southern Army and the IGHQ admitted the failure, and the Southern Army ordered the cancellation of the Imphal operation: 'The Imphal Area force shall move into a holding pattern.'[42]

The reason for the cancellation by the Imperial General Headquarters

The IGHQ reported the cancellation of the Imphal operation to the Emperor on 1 July 1944:[43]

> If the Imphal operation continues to be executed in the way we are doing now, our sound strategic system in Burma will collapse and, furthermore, there is the possibility that it could have a great influence on the conduct of the Great East Asian War. But if the Imphal operation were cancelled, then this would only have a small influence on our operations in Burma. But the mission to cut off the route between India and China is the most important component of our operations in Burma. If we abandoned this operation to cut off the route between India and China, our overall strategy would be greatly influenced because Allied air raids from Chinese bases against Japan as well as attacks targeting transportation between Japan and the Southern Area would take advantage of the cancellation of this operation.

The Imphal operation as executed under the Mutaguchi Plan, however, was not consistent with an operation designed to cut the Allied supply route between India and China. The objectives of the Japanese first offensive into Burma were to build a western wall to protect the southern resource area and to cut off the remaining outside routes supplying Chiang Kai-shek.

The primary objective of the Imphal operation should have been to protect all of Burma by attacking and destroying the enemy's supply bases in advance of an all-out enemy counter-attack. Mutaguchi had an ambitious dream to expel the British and liberate India if possible. Such overly ambitious dreams entailed enormous damage to Japanese forces and derailed the primary objectives of the operation, namely, to build a western wall protecting the self-sufficient Japanese co-prosperity zone and to cut off Chiang Kai-shek's last outside supply route. In August 1944, Myitkyina fell to the Allies, making the opening of the Ledo road a matter of time.[44]

From the beginning to the end of the Great East Asian War, the Japanese Army remained tied down in the China theatre. Was the Sino-Japanese War the cause of the Great East Asian War? The war against China was certainly an important factor. The imperative to cut off the Burma supply route to Chiang Kai-shek was clearly connected to the war in China. Yet the main purpose behind Operation Imphal was to create a defensive wall protecting the southern zone of Japan's co-prosperity sphere. Its cancellation occurred because its execution had precluded achievement of the operation's secondary objective of severing Chiang's links with the outside. The Allies made an enormous effort to supply Chiang in order to tie up as many Japanese troops as possible in the China theatre. With the Mutaguchi Plan, the Imperial Japanese Army played right into the Allied strategy.

Conclusion: Mutaguchi's war-leadership

Mutaguchi lacked two prerequisites for a successful strategy: battlefield intelligence and logistics. Wingate and his Chindits had helped inspire the Mutaguchi Plan, yet Mutaguchi ignored their vital air-supply network, which had made possible their advance. He was surprised by their ability to cross the Arakan Mountains and Chindwin River. So in his Imphal operation, he threw two divisions into the large mountain range and trackless jungles, giving little thought to supplies. He must have thought that the strong Imperial Japanese Army could cross this difficult terrain because the weak British force had already done so.

Mutaguchi based his optimistic assessment of enemy capabilities on his earlier experiences with the British forces retreating in Singapore. In planning his operation, he ignored the intervening changes and movements of the enemy forces. He was also indifferent to changes in the war and in technology. He should have studied the innovative British tactics of the 'admin box' or 'the cylinder position operation', that is the cubic (air–land) warfare during the Japanese second Akyab operation, executed just before the Imphal operation, but he believed in victory through flat (ground) encirclement. He believed the enemy would flee when threatened from the rear.

In the end Mutaguchi became transfixed by his own ambitions and unrealistic optimism. He stubbornly stuck to his original plans when he should have modified them in response to unanticipated enemy movements and changing logistical requirements. Success demanded the ability coolly to observe both the enemy's and his own actual situation. Exacerbating his short-comings was the growing tendency in the Imperial Japanese Army, regardless of feasibility, to reward positive, offensive war leadership and to penalise negative, defensive plans. This helps to account for the promotion of Lieutenant-General Mutaguchi from Commander of the 18th Division to Commander of the 15th Army. His promotion in 1943 to Commander of the 15th Army, charged with defending northern and central Burma, caused the tragedy at Imphal.

According to Yamamoto Tsunetomo's book on *Hagakure*, which Japanese military officers were very familiar with, 'If you really wish it, you can do anything.'[45] In this book Yamamoto argues over and over again that will-power is essential, not material superiority. It is no exaggeration to say that belief in the superiority of will-power was important in Japanese military thinking. But such ill-founded optimism was dangerous to Lieutenant-General Mutaguchi, Commander of 15th Army.

Mutaguchi considered himself to have been a key person in starting the Great East Asian War. And so he wanted to say it was he who had ended it. The Imphal operation, which was an ambitious attempt to end the Great East Asian War, resulted in the bitterest defeat on record for the Japanese Army. Indeed, it closed the curtain on the history of the Imperial Japanese Army as an institution. In this sense, Mutaguchi played an important role at the beginning and at the end of the Great East Asian War. In both cases, his role was negative.

9

BRITISH LEADERSHIP IN AIR
OPERATIONS

Malaya and Burma

Michael Dockrill

When General Douglas MacArthur, the US Supreme Commander of the South
Pacific Area, heard about Admiral Louis Mountbatten's appointment as
Supreme Commander, South-East Asia, at the end of August 1943, he urged
Air Vice-Marshal A. V. Goddard: 'Tell him that he will need more Air. And
when you have told him that tell him again from me that he will need *more
air.*' Here he thumped the table and almost shouted: 'And when you have
told him that for the second time, tell him from me for the third time that he
will need still MORE AIR.'[1]

MacArthur's warning had already been painfully learned by the British as a
result of their defeats in Malaya and Burma and elsewhere in the Far East in
1941 and 1942. In 1941 the then Commander-in-Chief, Far East, Air Marshal
Sir Robert Brooke-Popham, admitted that 'the strength of the Japanese Air
Force came as a complete surprise – in quality, performance and experience of
its personnel'.[2] The failure of the Royal Air Force could not be solely attributed
to the leadership qualities – or lack of – of its senior personnel. Both Brooke-
Popham, although elderly, and the ailing Air Vice-Marshal Pulford, the Air
Officer Commanding RAF, Far East, after 20 April 1941, were experienced
airmen, who did their best, in discouraging circumstances, to improve Malaya's
defences and to try, albeit unsuccessfully, to secure modern aircraft to replace
the Air Command's obsolescent planes. Air crews were inadequately trained,
radar was lacking, intelligence was poor, Brooke-Popham's authority over the
armed services was limited and there was considerable complacency about
the Japanese threat.[3] Brooke-Popham managed to secure 67 US Brewster
Buffalo aircraft for Singapore and Malaya by December 1941, but these were
underpowered and had a poor rate of climb compared with the far superior
Japanese Zeros. Otherwise Brooke-Popham had to make do with long-obsolete
torpedo bombers, the Vildebeests, a handful of Royal Australian Air Force
Hudson squadrons and four squadrons of Blenheim medium bombers. There

were no photographic reconnaissance, army co-operation or transport planes. The Japanese invaded Malaya on 8 December, destroying British aircraft on their airfields. They soon secured complete air superiority. Although 66 Hurricanes arrived at Singapore on 13 January 1942 they were much too late. The Japanese Air Force was steadily bombing Singapore's main airfields by day and night. The 127 operational Allied aircraft early in January were reduced to 71 by the middle of the month.[4]

On 27 December Brooke-Popham was replaced by General Sir Henry Pownall. Air Vice-Marshal P. C. Maltby arrived as Pownall's Chief of Staff. Then on 15 January 1942 American, British, Dutch and Australian Command (ABDACOM) was created with General Archibald Wavell as Commander-in-Chief, with Pownall now appointed as his Chief of Staff. Air Chief Marshal Sir Richard Peirse arrived from England to command the Allied air forces. Maltby established himself as Deputy Air Officer Commanding on 12 January 1945, leaving the desperately overworked Pulford in command. When the remaining Vildebeest squadrons were shot down in attempting to bomb an enemy invasion force near Singapore, what was left of the RAF was ordered to fly to Sumatra. With the fall of Singapore the Japanese now turned their attention to Burma.[5]

Slim commented that 'Burma was last on the priority list for aircraft, as everything else, and in December 1941 the air forces in Burma were almost negligible'.[6] When the new Air Officer Commanding, Burma, Air Vice-Marshal D. F. Stevenson, reached Rangoon on 1 January 1942 he found that he had only 21 P40 Tomahawk fighters and 16 Buffaloes under his command. Eventually, after much pressure, he was given 67 squadrons of Buffaloes, three squadrons of Hurricanes, many obsolete, and an American Volunteer Group, supplied with Tomahawks, under Major (later General) Claire Chennault. Stevenson's Command lacked adequate parts, proper repair facilities and effective anti-aircraft defences. Vigorous efforts were made by this relatively small air force to support the British Army in its retreat from Burma, inflicting severe losses on the Japanese Air Force before the fall of Rangoon on 6 March 1942. This victory enabled the British to evacuate Rangoon without substantial Japanese air interference. However, the Japanese Air Force soon rallied and destroyed the bulk of the RAF on the ground at Magwe air base. What was left of the RAF was withdrawn to India. By the end of May 1942 Burma had fallen to the Japanese.[7]

Apart from one disastrous excursion into Arakan in early 1944 British forces remained on the defensive in Assam while the British and US leaders argued about future strategy in the theatre. Churchill and the Chiefs of Staff wanted to bypass Burma altogether and produced interminable operational plans designed to recapture Singapore using amphibious forces to capture islands in the region, such as Sumatra, as a preliminary to the main assault. These plans were impractical and over-ambitious and they were finally abandoned when it became clear that no landing craft could be spared from North Africa or later from Overlord.[8] Furthermore the US Joint Chiefs, supported by Chiang Kai-shek and Stilwell, demanded that the British launch a major offensive to clear North Burma of the Japanese and reopen the Burma Road from Burma into China. Since the United States controlled the aircraft and the resources which were essential if the British were to embark successfully on any campaign in the theatre, the British had to give way. At the Quebec Conference in August 1943 the British Prime Minister, Winston Churchill, persuaded the United States to agree to the formation of a new Supreme Command for South-East Asia which would be separate from India Command. Churchill had no confidence in Wavell's military abilities, especially after Wavell's attempt to recapture the Akyab airfields on the Arakan coast earlier in the year had been repulsed with heavy losses. In fact Wavell himself had suggested earlier in 1943 that the Command be split as the existing theatre was too large for one overworked and tired man to cope with, especially given internal unrest in India and a serious famine in Bengal in 1943, both of which overtaxed the depleted air and ground forces in his Command. After various candidates had been suggested for the new Command – including Orde Wingate – Churchill appointed the relatively young (43) and energetic Lord Louis Mountbatten as Supreme Commander, South-East Asia Command. Mountbatten had been head of Combined Operations in England and Churchill was very impressed with his abilities, even though little in the way of Combined Operations had actually been carried out and the one that was – the assault on Dieppe – had been a disaster.[9] Mountbatten took up his new post in New Delhi in November 1943. Before his arrival British forces had faced defeat after defeat, and the only success, at least in helping to raise British morale, had been Operation Longcloth between February and May 1943, when British long-range penetration groups (the Chindits) led by Colonel Orde Wingate, penetrated deep into Burma in an attempt to destroy Japanese lines of communications. While a few rail and road connections were temporarily disrupted, Wingate pioneered the use of air power for his operations. Hitherto the British in Burma had been confined to the roads, which were easily cut and ambushed by the Japanese. Wingate used aircraft to carry his supplies – food, water, mail, mules, jeeps, guns, etc. – to pre-prepared landing fields or drop them by parachute. To further harry the enemy, fighter bombers carried out air strikes on Japanese positions. An RAF wireless section, commanded by Group Captain Robert Thompson, was attached to each of Wingate's columns to signal suitable dropping zones.[10] Pownall described Wingate as 'a genius in that he is quite a bit mad' and also as 'a thoroughly nasty piece of work'.[11] Louis Allen wrote that Wingate 'had changed the nature of jungle campaigning for good'.[12] Despite the relative paucity of his achievements, most accounts of his campaign agree that their audacity did a lot to raise British morale during a period of continuous defeats.

Mountbatten and indeed all the military leaders in India had long recognised that the key to the reconquest of Burma was a huge increase in the quantity and quality of British air power in the theatre – fighters, fighter bombers, and reconnaissance aircraft and above all air-supply and cargo-carrying planes. Despite Churchill's criticisms of the two generals – Auchinleck, he wrote, always demanded 'even greater force and . . . prescribe[d] far longer delays'[13] – Wavell as Commander-in-Chief, India and his successor after 20 June 1943, General Sir Claude Auchinleck, did order the construction of all-weather airfields in Assam, and they provided the infrastructure for Britain's subsequent victories.[14] The structure of the Allied air forces available to Mountbatten and General Sir William Slim, the Commander-in-Chief of Fourteenth Army, in December 1943 was a complex one. The Allied air forces and the one Dutch squadron in India and Ceylon had been commanded by Sir Richard Peirse since March 1942. He had been Air Officer Commanding, Bomber Command until 1941.[15] Sir Archibald Sinclair, the Secretary of State for Air, wrote to Churchill on 10 December 1941 that 'Peirse possesses not only recent experience in command, but also has a wide background of war knowledge and high staff experience. He is loyal, able and hard-working, and would be a strong support for C-in-C India.'[16] He was appointed as Commander-in-Chief of Air Command South-East Asia (ACSEA) on 16 November 1943. Although General Pownall, who had been appointed Mountbatten's Chief of Staff, thought Peirse 'definitely stupid',[17] Mountbatten was sufficiently impressed with Peirse to extend his appointment for six months in April 1944. The Supreme Commander wrote to Air Marshal Sir Charles Portal, the Chief of the British Air Staff, that he was on good terms with Peirse who 'has never failed to give me unbiased and fearless advice'. In September 1944 General G. E. Stratemeyer, Peirse's able US deputy, argued for a further extension of Peirse's appointment – he was 'a wonderful chap to work for'.[18] However Peirse's affair with Lady Auchinleck, Sir Claude's wife, whom Peirse eventually married, was now common knowledge and had led Peirse to neglect his duties. Mountbatten therefore refused a further extension and Peirse and Lady Auchinleck were flown to England on 28 November 1944.[19] He was replaced by Air Chief Marshal Trafford Leigh-Mallory, Air Officer Commanding in Chief Overlord, but, when the latter was killed in an air crash on his way to take up his command on 14 November, his place was taken by Air Chief Marshal Sir Keith Park on 24 February 1945, an inspiring leader who got on well with Mountbatten and the US personnel.[20]

As Air Commodore Henry Probert has pointed out, Peirse was a somewhat remote and aloof character but 'he fought interminable battles to win for his command the resources it needed for the Far East war; he led it through the multiplicity of operations . . . and he contributed most ably to the higher direction of the campaign in South-east Asia'.[21] In 1942 Peirse fought an energetic and ultimately successful battle with the Air Ministry about the reorganisation of his command structure, a reorganisation which the Air Ministry had initially rejected.[22]

He faced a gloomy situation. In May he informed Portal that 'Everything in India is unbelievably primitive. The totally inadequate staff and complete lack of things essential is quite devastating, and much of the personnel is past praying for – the Government of India is an Alice in Wonderland hierarchy.'[23] The air force in India consisted of a collection of obsolescent aircraft and its personnel was demoralised by repeated defeats. Because of India's inadequate industrial base, new aircraft, airmen, fuel and raw materials had to be imported via the Cape route, and took four months to reach India from Britain and the United States. However, despite enormous transportation problems and a lack of skilled engineers in India, by November 1943 275 airfields had been constructed in the north-east of India. Repair and maintenance facilities were also vastly improved, increasing the serviceability of aircraft to 80 per cent from 40 per cent in March 1942. Signal facilities, reconnaissance and air defences were also upgraded. By October 1943, although Hurricanes still formed the main defence force, high-performance Spitfires Mark VIII began to appear, while A1 Beaufighters were brought in from the Middle East for night fighter defence. Vengeance dive bombers were also used to assault enemy positions in Arakan.[24] The influx of new machines enabled Peirse, in his operational directives to his Command on 12 December 1943 and 21 January 1944, to set out the tasks of the Allied air forces. These were to protect the construction of a new road from Ledo to Myitkyina, which Stilwell's Chinese forces were in the process of constructing, to secure the air route from Assam to China and drive the Japanese Air Force out of Arakan. 3rd Tactical Air Force was to engage and destroy the enemy aircraft in the air, and attack enemy airfields and installations. Once land operations began, enemy targets in forward areas were to be attacked. Strategic Air Force was tasked with the mission of destroying Japanese shipping, factories, railways and bridges.[25]

However, Dakota transport planes, which Mountbatten and Slim were convinced were the key to future victory, were few in number. Britain depended on the United States for their supply. As early as November 1942 Wavell replied to an urgent request by General Irwin, GOC India, for more air transport planes: 'I wish I could get you more. They do not appear to be making any at all at home and we are entirely dependent on the Americans, who also appear to be in short supply.'[26] Pownall lamented in November 1943 that 'the problem of transport aircraft is a proper nightmare'. He found one full admiral and two full generals 'sitting round a table to try and screw out three Dakotas from somewhere – anywhere – to fulfil an urgent need'.[27] To make matters worse, 10th USAAF, which controlled most of Burma's transport aircraft, was not under Mountbatten's command but was a separate entity, commanded by General G. E. Stratemeyer, USAAF, since August 1943. He was responsible to General Joseph Stilwell, who had been appointed as Mountbatten's Deputy Commander but was also Chiang Kai-shek's Chief of Staff, an independent command based in Chungking. Most of 10th USAAF's transport aircraft were employed in ferrying supplies from India to Chiang's base in Chungking across 'the Hump'. The United States Joint Chiefs of Staff insisted on retaining control of all US aircraft in SEAC since they were intended to support the China theatre and not Burma. If Mountbatten wanted to divert planes from the airlift he had to ask the Joint Chiefs for permission through 'Vinegar Joe' Stilwell, who was an uncooperative and testy Anglophobe. All that Mountbatten had at his disposal for future campaigns in Burma were six squadrons of USAAF General W. F. Old's Troop Carrier Command, which had to meet the needs of both India and Burma. One of Mountbatten's first achievements as Supreme Commander was to persuade the US Army Chief of Staff, General George Marshall, and General Henry Arnold, the Commanding General of USAAF, at the Sextant Conference in Cairo in December 1943 to agree to the integration of the two air forces under Peirse, with Stratemeyer as his second in command. The integration took place on 12 December, despite strong opposition from Stilwell and Stratemeyer. The unified force became Eastern Air Command with Stratemeyer as its commander and at the time of its formation consisted of 735 aircraft: 464 RAF and 271 USAAF fighters, bombers and reconnaissance planes. The Command had a numerical superiority over the Japanese of about three to one.[28] Mountbatten commented that 'I absolutely had to integrate . . . the two Air Forces were running in watertight compartments with friction which occasionally came to a head.'[29]

The newly arrived Spitfires soon demonstrated their superiority over the Japanese Zeros when many Japanese fighters were shot down in an engagement above Chittagong on 31 December 1943 and along the Arakan coast on 15 January 1944.[30] After the arrival in Burma in early 1944 of US long-range P-51 (Mustang) and the P-38 (Lightning) fighters, Allied air supremacy was virtually assured. As a result Slim could now proceed with his planning for an offensive into Burma based on the setting up of strong points where British and Indian troops would remain if attacked by the Japanese and then, once they had defeated the Japanese, would resume their advance. Hitherto the Japanese tactics of encirclement and infiltration had successfully forced British forces to retreat, often in humiliating circumstances. These strong points were to be supplied by air, and Slim ordered his principal administrative staff officer, Major-General A. H. J. Snelling, to reinforce the air-supply units at Agartala and Comilla, intensify the training of the pilots and engage in the day-and-night packing of supplies to be airlifted to General Sir Philip Christison's XV Corps, which was to spearhead the invasion. To achieve these ends, Snelling worked closely with General Old and his Troop Carrier Command.[31] Christison began his advance to the south in November but on 4 February the Japanese opened up the Ha-go offensive designed to infiltrate, encircle and destroy the British forces in Arakan. This time, contrary to Japanese assumptions, the British did not retreat but held fast to their positions. Nevertheless the outcome of the initial battle of Ngakyedauk Pass in Arakan in early February 1944 was by no means certain. The Japanese deployed 34 fighters and ten bombers over the battlefield and initial Spitfire interception was not very successful. Furthermore attempts to air-supply Messervy's beleaguered 7th Division (the Battle of the Administrative Box) by 31 Squadron RAF had to be abandoned on 10 February. Seven out of 16 transport aircraft were lost to enemy fighters and light anti-aircraft and small arms fire from enemy-held areas over which the supply-dropping aircraft had to fly as low as 200 feet to reach the dropping zone. For a time air dropping by night had to be resorted to but after 14 February a fall-off in Japanese air activity led to the resumption of air dropping by day. The volume of supplies dropped sustained the besieged garrison until Slim brought up reinforcements and defeated the Japanese on 24 February.[32] In early March Christison was able to resume his offensive towards Akyab.

On 8 March 1944 the Japanese launched another and major offensive, U-go, which was intended to capture the British bases at Imphal and Kohima, as a preliminary to an advance on Chittagong. Mountbatten had already been convinced by numerous intelligence reports and aerial reconnaissance that the Japanese intended to strike at Imphal. Mountbatten was alarmed that neither General Sir George Giffard, Commander-in-Chief of 11th Army Group, nor Slim showed much urgency about reinforcing the Fourteenth Army in Imphal. It was Mountbatten who insisted on flying the 5th Indian Division from Arakan to Imphal, although the Official History states that Slim had already appealed for air transport as a matter of urgency. However, Troop Carrier Command was by now desperately overstretched and Mountbatten, ignoring Stilwell, informed the Chiefs of Staff that he was diverting 30 C-46 Commando and Dakota transports from US Air Transport Command operating over 'the Hump' – he had already, with the reluctant agreement of the Combined Chiefs of Staff, borrowed transports from 'the Hump' to help the air supply to Arakan, but these had been returned. On 17 February, after an appeal from Churchill, the US Chiefs agreed to the transfer but Mountbatten had already taken over the aircraft on the 15th.[33] Ronald Lewin commented that 'Mountbatten is to be seen making the big, right judgment and thus enabling Slim to concentrate on the battle'.[34] During the battle of Imphal the 5th Indian Division was moved from Arakan to Imphal complete with its guns, jeeps and mules between 19 and 29 March by Dakotas of 194 Transport Squadron RAF. The division's airlift has been described as 'a brilliant and successful improvisation which again demonstrated the confident flexibility of Slim's staff and the determination of the air crews'.[35]

Air Marshal Sir John Baldwin's 3rd Tactical Air Force was responsible for all air operations on the Fourteenth Army Front. He decentralised close air support to small air staffs attached to each group headquarters in order to make liaison with the Army more effective. On 1 May Troop Carrier Command was placed under Baldwin's authority, but deliveries to Imphal fell off in early May. Baldwin turned some airfields into single-commodity areas so that aircraft could be loaded to full capacity with one commodity. These measures led to a dramatic increase in supplies to Imphal during June, despite the huge strain on the air crews with difficult flying conditions and air crews operating to the limits of their endurance. Imphal Ground Control played a crucial role in ensuring that the airlift went as smoothly as possible under trying circumstances.[36] Earlier in the fighting at Imphal, General G. A. P. Scoones, Commander of IV Corps, suggested to the AOC of 221 Group, Air Vice-Marshal S. F. Vincent, based at Imphal, that he should move his units from the valley and operate from greater safety further back. Vincent refused, pointing out that he had done enough retreating from the Japanese and that he and his men would remain at Imphal where their presence was essential to the supply operation.[37]

Kohima was another crucial strategic position, situated as it was on the summit of the pass linking north-east India with Burma. In April the Japanese had besieged the Allied garrison there and the RAF had enormous problems in giving close air support and dropping supplies in the face of air pockets and sudden mists and the risk of being shot at from close range from Japanese positions close to the garrison.

Given the strain on the transport planes imposed by these operations, and the requirement to supply and support Wingate's second Chindit campaign, Operation Thursday, Slim's decision to fly part of General F. W. Messervy's 7th Indian Division from Arakan further to reinforce Imphal clearly required more transport planes. Accordingly, on 25 March, Mountbatten appealed to the Joint Chiefs of Staff for a further 70 Dakotas and the authority to retain the commandos he had taken from 'the Hump'. On 29 March the CCoS agreed to the temporary transfer to Mountbatten's command of 64 US and 25 RAF Dakotas from the Mediterranean and Middle East, as well as the retention of the planes from 'the Hump'. On 28 April, however, Washington demanded that the planes borrowed from the Middle East be returned by mid-May. This would have been disastrous for the Burma campaign, and Mountbatten faced an uphill struggle in his efforts to retain them. Although Washington offered to replace them by Dakotas from the United States, Mountbatten realised such replacements would be useless given that their inexperienced pilots would have had no training in the navigational and technical skills required in monsoon and mountainous conditions. He refused to let the Middle East planes go and eventually Washington relented and allowed him to retain the Middle East planes initially until 31 May and later until 16 June. Nevertheless, as Probert points out, supplies to Imphal were maintained by only a narrow margin.[38] The battles of Imphal and Kohima ended on 22 June when the British Army reopened land communications and destroyed the Japanese Army in the area.

Intensive Allied air operations over north Burma and attacks on Japanese airfields in southern Burma by USAAF General H. C. Davidson's Strategic Air Force equipped with Liberators (B.24) and Mitchells (B.25) achieved air superiority by the end of March, although in April Japanese fighters and bombers were able to resume daylight raids on Imphal but these were defeated by the RAF's adoption of air patrols which enabled the Spitfires to intercept the attackers, whose efforts steadily diminished over the next three months. The airlift of supplies to Imphal, Kohima and other threatened sectors by Eastern Air Command continued unimpeded by these Japanese attacks.[39] Of great importance during the campaign was that non-combatants and the sick and wounded could be flown out of Imphal and Kohima. Both Mountbatten and Slim insisted that this was essential if the spread of disease in the garrisons was to be contained. Slim wrote: 'air evacuation, in the long run, probably made the greatest difference of all to the wounded and sick'.[40]

Meanwhile Wingate's special forces had landed deep in the interior of Burma to harass Japanese communications, especially the Mandalay–Myitkyina railway, by establishing strong points in occupied territory – the latter likely to deprive the Chindits of the mobility which had been the key to what they had achieved in Chindit 1. The need to fly in the bulk of the Chindits and keep them supplied once they had landed strained the air transport resources of SEAC to the maximum. Wingate demanded more Dakotas to support his operations but Slim rejected his request – they could not be spared. Wingate was killed in an air crash on 24 March 1944. Military historians and soldiers continue to argue about his achievements – Slim wrote later: '. . . I do not believe that the contribution of Special Forces was either great in effect or commensurate with the resources it absorbed'[41] – but there is no doubt that Wingate was an inspiring leader and most of his men were devoted to him.[42] After his death the Chindits' operations continued, latterly in support of the advance of Stilwell and the Chinese forces who succeeded in capturing Myitkyina airfield on 7 May 1944, again using air transport to sustain the campaign.[43]

General Sir Montagu Stopford's XXXIII Corps now began to advance towards the Chindwin in monsoon conditions. Slim hoped to use the Chindwin as a base for an advance on Mandalay and Rangoon in the next dry season. However, General A. C. Wedemeyer, Mountbatten's former Deputy Commander, who had been transferred to Chungking to take over from Stilwell after the latter had been relieved of his command in October 1944, suddenly ordered the diversion of three squadrons of Dakotas (75 planes) in December to airlift Chinese divisions to China to defend Chennault's 14th Air Force airfields from a Japanese offensive. At first Mountbatten's protests were overruled by the Chiefs of Staff, but, eventually, after much pressure from Mountbatten, Wedemeyer, on 1 February 1945, returned two of the squadrons while the British diverted transport planes from other theatres to Burma. Finally Mountbatten persuaded the Combined Chiefs of Staff to let him retain the two squadrons until 1 June or until the fall of Rangoon, 'whichever date is earlier'. This was crucial since Slim's advance on Mandalay, which fell to his forces on 9 March, and then on Rangoon, depended totally on air supply.[44] Vincent reported to Mountbatten on 23 February 1945 that the close tactical air support which his squadrons provided had made up for the shortage of artillery which the terrain made difficult to move up. He continued: 'Almost every aeroplane exceeded the maximum degree [of operational intensity laid down by the Air Ministry] for *six consecutive months* – a wonderful effort of the pilots and the ground crews and also of the aircraft depots back in India which kept us supplied with remarkably little waiting.'[45] The aircraft dropped not only food, ammunition, fodder and water but also mail, drugs, boots, the SEAC newspaper, typewriter ribbons, socks, toothbrushes, razors, spectacles and even rum.[46] Rangoon, which the Japanese had evacuated, fell to the Allies on 2 May 1945. Apart from mopping-up operations, many of which still involved ferocious air and land fighting, that was the end of the campaign – the Japanese armed forces surrendered to the British in Singapore on 12 September 1945 (they had already formally surrendered to MacArthur at Tokyo on 2 September). The Japanese collapse came at a fortunate time since it appeared that the Air Ministry was more anxious for the Royal Air Force to participate in the US bombing campaign against Japan than in providing up-to-date bombers to pave the way for Mountbatten's invasion of Malaya.

Although, as Ronald Lewin points out, Mountbatten was 'not free from error',[47] the leadership qualities of Mountbatten during the Burma campaign were not in doubt, especially in his overall responsibility for air. Despite his criticisms of Mountbatten's occasional impulsiveness and failure to consult staff before issuing directives, Pownall wrote on 14 September 1943 that 'Mountbatten, aged 43, will certainly have all the necessary drive and initiative to conduct this war'.[48] In November Pownall found that 'his energy and drive are most admirable features'.[49] He certainly galvanised the lethargic Indian organisations, and his most important contribution to victory was in securing from the United States much needed transport planes in the face of US obstruction. Stratemeyer, who loyally supported Mountbatten during the Burma campaign, described him as 'An outstanding example of how an Allied Air Commander should conduct himself'. Andrew Gilchrist of the British Foreign Office, who was on active service in South-East Asia between 1944 and 1945, wrote: 'When I think of the handicaps under which Mountbatten laboured – repeated hold-ups of promised man-power supplies and shipping, fantastic political entanglements due to American policy in China – I am amazed that so much was done.'[50] Of the other air leaders, Peirse's deputy, Sir Guy Garrod, who took over the Command temporarily until the arrival of Park in late February, was responsible for the planning of air operations which contributed so crucially to the Allied victories in Burma in early 1945. Stratemeyer described Garrod as 'a natural commander and leader . . . he's a great guy'.[51] This compliment could also be applied to Air Vice-Marshal S. F. Vincent who led 221 Group at Imphal with conspicuous success. Air Commodore the Earl of Brandon, who took over 224 Group in July 1944, was highly respected by his men and by other service commanders, although his unorthodox ways frequently upset senior officers. In the struggle over Mandalay he removed his air rank badges and flew operational sorties with 273 Squadron as a flying officer.[52] Mountbatten also received the full support of General Raymond Wheeler, US Army, his principal administrative officer and an expert on logistics. Mention should also be made of Air Vice-Marshal T. M. Williams, who replaced Stevenson as AOC Bengal in January 1943. Slim described

Williams as 'an inspiring commander . . . who laid the foundations of the air supremacy we subsequently gained'.[53] Another daring leader was USAAF General Old of Troop Carrier Command who often led sorties himself.[54]

Despite Pownall's complaint in April 1944 that the Strategic Air Force was attacking Japanese lines of communication instead of their more crucial aerodromes and shipping and that 'It's the old game of "the separate war" of the Air Force',[55] the record shows that, for the most part, co-operation between air and ground commanders was close and cordial, as was co-operation between USAAF and RAF units. This was largely due to the leadership qualities of both Slim and Baldwin who worked closely together at Comilla earlier in the campaign. Nor should be forgotten the tenacity and courage of air crews throughout the campaign who frequently flew to the limits of their endurance to sustain the ground campaign. They were rightly praised by Sir George Giffard in a letter to Slim on 28 July:

> I have not forgotten the immense debt which the Army owed to the Air. It is no exaggeration to say that without the magnificent assistance given by the Eastern Air Command, the Army could never have won its victories.
>
> I am sure that no one who watched them is ever likely to forget the courage, determination and skill of all the aircraft pilots and crews who have flown in the worst weather in the world over appalling country either to attack the enemy in front of the Army and his communications in the rear with bombs and machine guns or to deliver reinforcements, supplies, ammunition etc to the troops isolated in Arakan, Imphal and Central Burma.[56]

10

AIR OPERATIONAL LEADERSHIP IN THE SOUTHERN FRONT

Imperial Army Aviation's trial to be an 'air force' in the Malaya offensive air operation

Hisayuki Yokoyama

Introduction

The Imperial Japanese Navy's use of air power in the role of a 'naval air force' during the Second World War has been well studied. In fact, the air attack on Pearl Harbor and the subsequent naval battle off Malaya Peninsula are recognised as the key turning points that pushed air power toward centre stage in naval battles, replacing the traditional battleship-oriented naval power philosophy that had dominated the Imperial Navy since its birth. In addition, we can see a similar change and many signs and traces of hardship on the transformation within the Imperial Army. During the offensive air operation over Malaya, Imperial Army Aviation tried to be an 'independent army air force' growing out of the traditional concept of operations that emphasised supporting ground forces by reconnaissance and close air support, etc. When the Malaya offensive air operation started, Imperial Army air power could only reach northern Malaya owing to the insufficient combat radius of aircraft operating from original deployment airfields in southern French Indo-China. However, within only a few days, Imperial Army Aviation gained almost complete air superiority over Malaya utilising the tactics of 'aerial exterminating action' and occupying airfields one by one in the Malaya peninsula. This initial successful operation significantly contributed to the subsequent Imperial Army land campaigns in Singapore, Java, Sumatra and Burma. The Japanese official war history series praised these Imperial Army Aviation tactics as a historic precedent that was equal to the Imperial Navy's Hawaii operation.[1] The British official war history series, too, recognised the significance of the early loss of air supremacy and stated that 'before the war was two days old, the situation of the Royal Air Force in northern Malaya, and therefore of the country as a whole, always weak, had become gravely compromised'.[2]

Two leaders played particularly significant roles in the success of the Malaya air operation, Colonel Kazuo Tanigawa and Lieutenant-General Michioho Sugawara. Tanigawa, an air staff member of the Southern General Army Headquarters, planned the Malaya offensive air operation. Sugawara, Commander of the 3rd Air Corps, led this operation. The operational concept they both considered ideally suitable and indispensable was Army Aviation as an independent 'air force'. They emphasised the gain of control of the air by 'aerial exterminating action', and considered support to ground forces and strategic bombardment as secondary.

The purpose of this chapter is to argue that Imperial Army Aviation took large steps toward becoming an 'air force' in the Malaya offensive air operation, and to highlight the leadership of Tanigawa and Sugawara. This essay will first show how the 'aerial exterminating action' concept was adopted as the operation of an independent 'air force' along with the development of other air operational concepts in Imperial Army Aviation. Secondly, this essay will discuss how Tanigawa tried to utilise air operations in an 'air force' role as he met with resistance from the traditional ground force thinking within the Imperial Army. Thirdly, it will consider how Sugawara's operational leadership is shown in the Malaya offensive air operation. Finally, this chapter will mention some important weaknesses in Tanigawa and Sugawara's operational concept of an 'air force'.

The terminology of 'aerial exterminating action' in the Imperial Army meant positive and autonomous surveillance and complete destruction of enemy aircraft in the air and while still on the ground at their airfields. Thus the term is considered the equivalent of what the Japanese air defences refer to today as 'offensive counterair' (i.e. Japanese air offensive action against all elements of the enemy's air force both in the air and on the ground). For serious application of 'aerial exterminating action', doing interception and pursuit all the time was strictly frowned upon in Imperial Army Aviation.[3]

Introduction of the air operations concept as an 'air force' and the adoption of 'aerial exterminating action'

The origin of Imperial Army Aviation dates back to the formation of the Qingdao dispatch squadron that was organised provisionally to participate in the First World War to capture the Qingdao in August 1914. The following year, an air battalion consisting of two flying squadrons and one balloon squadron was formed. By 1923, a total of four air battalions had been established. In the early stages of Imperial Army Aviation, reconnaissance aircraft and fighters were expected only to support ground forces as airborne eyes and longer hands. Reconnaissance flights flew to observe enemy activities, report impact points to artillery groups, conduct liaison flights and perform an occasional bombing mission. Fighter flights sometimes guaranteed freedom for friendly reconnaissance flights in the battlefield and occasionally disrupted the enemy's reconnaissance flights, but, in spite of this fighter contribution, the primary mission assigned air power in those days was reconnaissance.[4]

The first step toward an independent 'air force' was marked in 1925 by the transformation of the Imperial Army resulting from 'Ugaki's disarmament'. By abolishing four infantry divisions, this disarmament freed funds to update equipment and was intended to convert reconnaissance-oriented air power to a bomber air power capability.[5] With subsequent reinforcement, the bomber force achieved an equivalent force ratio with that of fighters in 1936. During this same period, Western countries were trying to determine the best methods and most appropriate uses for military air power. Italian General Giulio Douhet emphasised the importance of bombardment units in future air battles in his 1921 book, *The Command of the Air*, and US Army General William E. Mitchell advocated a strong, independent air force for air power in the United States. Even before Douhet's book was published, Great Britain had taken the lead in establishing an independent air force for defensive counterair against air strikes from the European Continent. But there was no common clear answer in the United States or Europe about what the emphasis of air power should be, supporting ground forces or performing air strike and strategic bombing as an 'air force'. Meanwhile in Japan, the Imperial Japanese Army in 1919 accepted an offer from France to receive training from the air tactics education team led by French Army Colonel Jacques Paul Faure. This so-called 'Faure Wing' taught young Japanese aviators basic flying, aerial firing, reconnaissance and bombing. The report on the outcome of this training emphasised the importance of bombing in both daytime and night-time by an independent bomber force.[6] However, the Imperial Army was not equipped with bombers. Bomber units were not organised until 'Ugaki's disarmament'. Major reasons for the delay were Faure's influence as an opponent of the concept of an independent 'air force', and the operational concept of ground support deeply rooted in the Imperial Army. Though the character of their air power changed to include a bomber force after 'Ugaki's disarmament', the dominant operational concept within Imperial Army Aviation remained supporting the ground forces as their primary role.

However, there were winds of change blowing with regard to the operational concepts within Imperial Army Aviation in 1934. Concepts were introduced to attach more importance to offensive counterair than to a strictly ground support-centred mission. For the first time, the Imperial Army adopted the idea of 'aerial exterminating action' in the 'Airman Drill Book' compiled as an airbattle standard for air units. The manual still placed reconnaissance and close air support as the main missions of air units, but prescribed that 'aerial exterminating action' must be the means to assist ground forces to reach a successful position.[7] Also, the manual referred to the importance of control of the air to ensure favourable progress for the entire operation. The reference to control of the air was the beginning of an 'air force'. The concept to gain control of the air by 'aerial exterminating action' was totally adopted as a primary tactic in the Annual Operation Plan of 1935. At that time, the Imperial Army was aware of air power reinforcement by the Soviet Union in the maritime province of Siberia. The Imperial Army's estimate of Soviet reinforcement was that Soviet heavy bomber units were capable of attacks against Manchuria and the Korean peninsula, the Japanese mainland and perhaps even Tokyo. Therefore the Imperial Army began to study possibilities for 'aerial exterminating action' to destroy Soviet heavy bomber capability in the maritime province quickly at the beginning of a war. The concept of this study was adopted into the Annual Operation Plan of 1935.[8] Imperial Army Captain Shimanuki Takehar, who was responsible for drafting this plan, recalled this trend toward 'aerial exterminating action' in his memoirs. He wrote that strategists within Imperial Army Aviation believed the best use of air power was 'aerial exterminating action' to achieve success by control of the air.[9] Certainly, as Simanuki said, the idea of air operations to gain control of the air became prevalent within Imperial Army Aviation at that time. However the basic concept of operations was still the ground support mission. 'Aerial exterminating action' was intended as means to gain an advantage in the whole of Army operations. 'Aerial exterminating action' for gaining control of the air was to be limited to the beginning of a war in this Annual Operation Plan. In the next phase, concentrating on supporting ground forces was necessary. This was because Imperial Army Aviation could not accomplish the destruction of the overwhelming Soviet air power threat in the Far East at such a great distance. As a means of opposing the Soviet air threat, the Imperial Army only planned to use 'aerial exterminating action' in the first phase of a war.

The above dominant operational concept changed when the Army Air Service Head Office compiled 'Air Troops Operations' as a norm for leadership of air units in December 1937. This document stated that the main object of air operation was to destroy enemy air power. The result of 'aerial exterminating action' immediately after the outbreak of a war, especially first strike, held the key to success of subsequent air operations, and it had a great influence on the whole of the operation. But it also stated that close air support and strategic bombing would be executed based on the circumstances and at the proper time.[10] This document was drafted to show operational standards for air units to form the basis for long-term acquisitions of air power assets. This programme was started in 1934, and was in its final stages when 'Air Troops Operations' was completed. This programme emphasised bomber units with the component ratio of air units as five-to-three-to-two of bomber, fighter and reconnaissance assets, respectively.[11]

The idea of 'Air Troops Operations' was greatly influenced by the report of the Ohshima Air System Inspection Delegation to Germany. The delegation, headed by Major-General Hiroshi Ohshima, a military attaché to Germany, visited the German Air Force in order to help develop the long-term arms programme. Sugawara, who was a colonel and Chief of the First Section in the Army Air Service Head Office, was assistant chief of this delegation.[12] In his report, Sugawara applauded Great Britain's achievement of an independent air force, and Britain's aim to reinforce its air power as it foresaw how the strategic progress of air power might mitigate the defensive capability of an island nation. Sugawara appealed that Japan must build up an independent great air force too, because Japan was in almost the same strategic situation. He also emphasised the importance of the destruction of enemy air power by pre-emptive raids at the beginning of a war. Although he admitted the importance of supporting ground operations in times of decisive battle, he felt it unwise to restrain air units throughout the whole period of ground operations because of the necessity for a broader application of air power throughout a war.[13]

'Air Troops Operations' was treated as no more than a mere study aid by Army leadership because the concept of an independent 'air force' with the means of 'aerial exterminating action' slighted the Army leadership's aviation concept of primary support to ground operations. Consequently, 'Air Operation Essentials' was authorised as the operational standard in 1940. This manual reflected lessons from the Nomonhan Incident,[14] and treated the importance of supporting ground operation and that of 'aerial exterminating action' equally to neutralise any inclination toward 'aerial exterminating action'.[15]

Tanigawa's attempt to utilise Imperial Army Aviation as an 'air force' in the Malaya offensive

The process of planning the Malaya offensive air operation and the awareness of control of the air

By curious coincidence, the Imperial Army and Navy started to plan the offensive operation of strategic vital areas such as Hong Kong, Singapore, Borneo and Malaya, assuming Great Britain as a hypothetical enemy when 'Air Troops Operations' was compiled. At first, this offensive operation was the naval plan. But the Imperial Army took responsibility after the Annual Operation Plan of 1939, and saw difficulties with accomplishing the plan.[16] Although the operational concept of Imperial Army Aviation remained the support of ground operations at that time, they realised from current wisdom that air strikes were indispensable at the beginning of a war. However, they did not have the striking range necessary to conduct 'aerial exterminating action' before the landing operation in the Malaya offensive operation. To begin with, Type 97 fighters (Ki-27) of the Imperial Army's mainstay had only 400 kilometres of combat radius. Therefore, it was not possible to escort amphibious squadrons the distance of 600 kilometres across the Gulf of Thailand, even if they could hold some airfields in southern French Indo-China. Secondly, even Type 97 heavy bombers (Ki-21) could only reach northern Malaya.[17]

Eventually, they tried to overcome their difficulties by using tactics where advance elements executed surprise landings and capture of airfields before 'aerial exterminating action'. This idea was based on the result of a map manoeuvre exercise at the Army Staff College in January 1941. The purpose of this manoeuvre exercise was to evaluate the plan drafted by Lieutenant-Colonel Tanigawa Kazuo of Military General Staff. Participants were Colonel Miyoshi Yasuyuki, Instructor of Army Staff College, as director of this manoeuvre exercise, and students of Army Aviation as exercise players. Before he wrote the plan, Tanigawa spent nearly one month in 1940 investigating British Malaya, French Indo-China and Thailand for operations against British forces of the southern area. From this experience, he realised the most difficult problem for the execution of this operation was the lack of combat range of Imperial Army fighters.[18] To address this problem, the Imperial Army speeded up the process to adopt the Type 1 fighter (K-43) with a longer combat radius. This new fighter was to have a combat radius of from 600 to 700 kilometres using the same engine as the Imperial Navy Type 0 fighter (A6M). The Type 1 fighter enabled the following: air cover of an anchorage area by fighters, and 'aerial exterminating action' to northern Malaya from southern French Indo-China by a strike package of bombers and fighters. Based on using the Type 1 fighter, Tanigawa devised the original plan where a section of ground forces would begin to land in northern Malaya to capture airfields. After 'aerial exterminating action', the ground forces' main body would land in Mersing, in north-east Singapore, with the support carrier group. But Director Miyoshi set the exercise to land at Singora and then go down through the Malay peninsula. He thought Tanigawa's plan would not guarantee enough control of the air for this operation because he could not count on support from the carrier group at that time.[19] In fact, the Type 1 fighters available for use in the Malaya offensive air operation consisted of only two regiments (60 fighters). The difference in thought came from their operational concept. Tanigawa made much of 'aerial exterminating action' at the beginning of a war to gain control of the air. However, Miyoshi considered the long theatre of the Malaya peninsula and instead placed greater importance on ensuring control of the air step by step by gradual advance.

Next is an example of how they actually gained and maintained control of the air in the Malaya offensive air operation. The Imperial Army attached much importance to the speed of offensive action in both the ground and the air operation in Malaya. Imperial Army Aviation perceived that this speed was restricted by the speed of gaining control of the air, and the advance of control of the air was related to the speed of capturing airfields.[20] Imperial Army Aviation adopted tactics to accelerate control of the air by utilising 'aerial exterminating action' and capturing airfields for Imperial Army Aviation to use one by one to complement the still-short combat radius of the fighters. In the opening battle that launched this operation, the 5th Division, as advanced elements of the 25th Army, started surprise landings in Singora and Patani.

The Takumi Column, commanded by Major-General Takumi Hiroshi, and the Uno Column, led by Colonel Uno Misao, landed in Kota Bharu and Nakhotn, Bandon, before the start of 'aerial exterminating action'. These landing forces included air section troops such as aircraft maintenance crew, crew for repair and management of airfields, etc. The landing of the Takumi Column was at 2.15 a.m. on 8 December 1941, almost one hour before the attack on Pearl Harbor. 'Aerial exterminating action' against the airfields at Sungei Petani and Alor Star in Kedah province began at 8.20 a.m., almost six hours after the Takumi Column landing.

Miyoshi's plan included a traditional area of conflict in the operational concepts between ground forces and Army Aviation. This conflict was the order of priority between 'aerial exterminating action' for gaining control of the air, and the ground support mission. In his plan, the speed of gaining and expanding control of the air would be subject to the speed of the capture of airfields, where the capture of airfields was a ground operation itself. This was the dilemma of Tanikawa. He longed for air operations as an independent 'air force' in Malaya.

The endeavor to carry out independent air operations in Malaya

After the map manoeuvre exercise at the Army Staff College, Tanigawa began to devise a full-dress plan of offensive air operations while serving as a staff member of the Army Division of the Imperial General Headquarters from early September 1941. At that time, the 'Outline of Imperial Policy' to assume the offensive in the Southern area was decided. At the beginning, the operations area of the Southern area was divided into the Philippines front and the Malaya/Singapore front. After Tanigawa completed the plan, he became the chief of the air staff office at the Southern General Army Headquarters on 13 November 1941. There is evidence showing the essence of Tanigawa's operational concept as an 'independent army air force' in his plan and leadership of the Malaya offensive air operation. This is seen in the concept of 'independence of aviation units' and 'uniqeness of air operations'.

Looking first at 'independence of aviation units', Tanigawa tried to gain independence of aviation by giving discretionary powers concerning air operations to the Southern General Army in the Southern offensive campaigns plan of the Army Division of the Imperial General Headquarters. His campaign plan to put whole aviation units under the direct control of the Southern General Army Commander was approved on 5 November. In his plan this operation would initially begin by the landing of advance elements in Malaya and by air attack on the Philippines. Next, major elements of the ground forces would land in the Philippines and Malaya and should quickly capture them based on a successful result of rudimentary air operations. Based on the guidelines of this air operation, Imperial Army Aviation units were expected to gain control of the air by pre-emptive attack on enemy airfields jointly with Imperial Naval air units. This was intended to steer landing operations of major elements to success followed by air support to ground operations. The main body of Imperial Army Aviation advanced to Malaya. The Southern General Army was organised with the 14th Army, primarily consisting of two infantry divisions for the Philippines offensive operation, and the 25th Army of four infantry divisions for Malaya, as the direct control units; and the 3rd Air Corps of three air brigades, the 5th Air Corp of two air brigades and other small air units.[21]

This plan gave priority to the ground support mission, but Tanigawa, as air staff officer of the Southern General Army, took direct control of aviation units to pursue the plan for independent air operations. In the plan of Southern General Army, Tanigawa attached the 5th Air Corps to the Commander of the 14th Army, but he left the 3rd Air Corps under direct control of Southern General Army. By not attaching the 3rd Air Corps to the 25th Army, the organisation of the task force placed the 3rd Air Corps on an equal basis with the 25th Army. In addition, he reinforced the 3rd Air Corps by attaching one air brigade of the 5th Air Corps. The 3rd Air Corps was composed of the 3rd, 7th 10th and 12th Air Brigades. Thus the 3rd Air Corps had a total of 612 aircraft, and was the main force of Imperial Army Aviation.[22] This source of air power was used to capture Malaya as the main air operation of the Southern General Army. The reason for this quantity of aircraft was based on the following estimate. The Imperial Army projected that British Air Force aircraft were around 200 to 250 in Malaya, 200 in India and 50 in Burma. The Imperial Army planned to deploy two or three times that number to achieve superior air power potential in the main theatre.[23]

The issue concerning attachment of air troops was the most difficult and anguishing issue for Tanigawa. He finally came to a conclusion that did not attach air troops to ground forces, but had air troops supporting the ground forces. He had anxiety about the ground forces' lack of understanding concerning the uniqueness of air operations, and he was furthermore apprehensive about possible obstructions to progress in overall Army operations by the feud within the Imperial Army between 'aerial exterminating action' and the ground support mission that had existed since the compilation of 'Air Troops Operations'. Tanigawa reasoned that the 3rd Air Corps, who supported not only the 25th Army of Malaya but also the 15th Army of Burma, might have a chance to shift the centre of gravity of 'aerial exterminating action' depending on the situation. He also felt that the ground forces did not have staff with experience and expertise to fully realise the complex mechanism and operations of air power operations. Above all, Tanigawa feared that lack of understanding by ground forces could bring about an unnecessary loss of valuable and scarce air assets.[24] Tanigawa thought that control of the air by 'aerial exterminating action' contributed more to the overall army operation than just providing ground support mission. He tried to plan air operations mainly to gain control of the air under direct control of the Southern General Army, but Tanigawa understood that attachment to ground forces was also favourable for acquiring enemy airfields.

This issue became apparent during the Malaya offensive operation. The 83rd Independent Flying Regiment was to be attached to the 25th Army in the plan, but deployment of this regiment was delayed at the beginning of the operation. Therefore the staff office of the 25th Army was concerned that there was not enough air support for the landing operation of advance elements and requested the attachment of flying units to provide ground support for exclusive use. Thus, emotional disagreements occurred between the 25th Army and the 3rd Air Corps. A certain air staff officer of Imperial General Headquarters who visited front-line ground forces on 19 December suggested that the 3rd Air Brigade be attached to the 25th Army temporarily.[25] Tanigawa also investigated the situation, and pointed out that Lieutenant-General Tomoyuki Yamashita, Commander of the 25th Army, did not always want the attachment. The misgiving of the 25th Army's staff office about this attachment came from an incomplete linkage between air and ground operations during the previous week. He observed the essence of this issue as follows. The 25th Army only discussed whether or not flying units directly supported ground forces without considering the profit from gaining control of the air. The staff of the ground forces did not understand the complex mechanism of flying units.[26] Tanigawa eventually refused the request of the 25th Army, and kept the 3rd Air Corps under direct control of Southern General Army to maintain the independence of aviation units.

Secondly, we examine the 'uniqueness of air operations'. In the operation plan of the Southern General Army, considering the characteristics of the Southern theatre, Tanigawa gave a free hand to the 3rd Air Corps on air operations under direct control of the Southern General Army, and tried to facilitate the unique nature of air operations without having them being dragged into exclusive ground operations. In the first phase of this operation plan, the mission of the 3rd Air Corps was mainly exterminating enemy air power and supporting the 25th Army's operation using the 3rd Air Corps major elements. Next the 3rd Air Corps supported the 16th Army in the Southern Sumatra operation by trying to cut enemy communication lines and by strategic bombardment. In the second phase, the mission was supporting the 16th Army in the Javanese theatre by a large number of the 3rd Air Corps elements supporting the 25th Army. Tanigawa gave the 3rd Air Corps the primary mission of exterminating enemy air power. This is because it was impossible for the 3rd Air Corps to be exterminating and supporting at the same time while air and ground operations moved simultaneously in the first phase. It seems that the writing together of exterminating and supporting in the Imperial Army Aviation was only a compromise to the conflicts between aviation and ground forces in Imperial Army. In addition, this air operation was planned to 'exterminate' and 'support' in the vast theatre from Burma to Java.[27]

It seems quite natural and reasonable that gaining control of the air should have been the main mission of this air operation in its contribution to the overall campaign.

For instance, 'aerial exterminating action' to Rangoon was executed twice on 23 December during the Malaya operation. This operation was in order to eliminate the enemy air threat and to apply political pressure by air attack to the heart of Burma as British air activities became conspicuous in Burma. But this air operation ended with an unexpected loss because of insufficient co-ordination between bombers and fighters. In another example, the Singapore operation and the Palembang airborne operation were conducted at almost same time. The 3rd Air Corps launched 'aerial exterminating action' on Palembang on 6 February 1942, and supported the Singapore landing operation by the 25th Army from 9 February. The fall of the Singapore fortress was 15 December, and the airborne assault on Palembang was 14 December, the previous day.[28] The Commander of the 3rd Air Corps also led paratroops at that time. This shows how Imperial Army Aviation demonstrated the uniqueness of operations through 'aerial exterminating action', ground support operations and airborne operations.

Tanigawa developed planning and operational leadership from a perspective of the 'independence of aviation units' and the 'uniqueness of air operations' in the Malaya offensive operation. It was his goal to execute air operations as an 'air force' for gaining control of the air by 'aerial exterminating action' during actual combat operations.

Leadership of General Sugawara, Commander of the 3rd air corps, in the Malaya air operation

The operational concept of Sugawara

The operational concept of Sugawara on air power was to put first priority on gaining control of the air through 'aerial exterminating action'. His thought was clearly shown in the report of the Air System Investigation Delegation to Germany. There were two aspects to this concept. First was to emphasise the execution of pre-emptive attack by 'aerial exterminating action' at the beginning of hostilities and to concentrate on air power at that time. Second was to carry out the operation as an independent 'air force' by having Imperial Army Aviation strike enemy air power using 'aerial exterminating action'.[29]

His emphasis on 'aerial exterminating action' at the beginning of hostilities was evident when he criticised the operations plan developed by the Army Division of the Imperial General Headquarters. In October 1941, Major Takagi Sakuyuki, the staff officer of the Imperial General Headquarters, notified Sugawara of a change of course of action on the Malaya offensive operation where a surprise attack landing would precede 'aerial exterminating action'. Sugawara wrote in his diary that he could not agree with this concept of operations, and that the intention of the Imperial General Headquarters was hard to understand. Sugawara thought it was of no use discussing this with the dispatched staff officer, but then Sugawara could not fall asleep.[30] Later as Takagi looked back on this encounter he recalled that Sugawara must have been apprehensive about a landing operation without 'aerial exterminating action'.[31] This confirmed that Sugawara had apprehensions about a landing operation without 'aerial exterminating action', as Takagi thought. However, above all, Sugawara worried that if 'aerial exterminating action' could not be 'pre-emptive', especially in consideration of timing, then it might be hard to achieve a successful result at the beginning of an operation. Sugawara ultimately arrived at the conclusion that if the word 'pre-emptive' remained in the operations plan, a 'pre-emptive raid' could include 'aerial exterminating action' so that 'aerial exterminating action' might take place even without ground forces landing operations.[32]

For example, when Imperial Navy aircraft found a British seaplane and shot it down on 7 December, Sugawara thought it might be a chance to start an offensive operation, but he gave up. What he worried about most in the beginning of air operations was how to ensure a 'pre-emptive' attack at the outbreak of a war. He thought that, without a positive offensive operation, the intended air operations would be less successful and perhaps fail from the beginning. Once the air operations had failed, Imperial Army forces might suffer irrecoverable damage in a short time. Then it would not be possible for air operations to reconstitute and change to a defensive posture.[33]

Next looking at independent operations as an 'air force' by Imperial Army Aviation, Sugawara proposed an independent great air force in the report of the Air System Inspection Delegation to Germany. At that time, as the precondition to establishing an air force, he insisted that Imperial Army Aviation should be independent in the Imperial Army structure itself. But he understood that independent operations were not so easy from the tough discussions among the Army Staff as to whether or not the Imperial Army Aviation Candidate School was to be a separate training programme. However, in spite of the dominant atmosphere within the Imperial Army advocating that air power must be a part of the ground forces, Sugawara tried to command his Air Corps as an 'air force' in the Malaya offensive air operation.

Concerning the attachment and support to ground forces issue about which Tanagawa made his strenuous effort, Sugawara interpreted this as a debate originating from the so-called 'reconnaissance era'. He strongly believed that during the evolutional development of air power it must be taken for granted that support to ground forces should be done as 'collaboration' with ground forces. The debate itself stemmed from the planning of the Southern Offensive Operation Plan, but Sugawara explained that it was only a matter of feeling. He asserted there to be no critical loss of support to ground units because an air unit that works hard under the command of ground forces would surely accomplish its mission even in 'collaboration' status. He explained that, unlike with artillery units, the attachment of air units does not mean 'direct command'. This is because they must be apart from each other geographically. Because of this separation it would likely be difficult for ground forces to understand the logistics for air power, aircraft maintenance, a supply of fuel and ammunitions, and so on. He saw that the frustration of ground forces was not an operational support issue, but rather a preoccupation that air units should be attached to ground forces just to follow the tradition of the Army in the 'reconnaissance era'.[34] Sugawara wrote in the last part of his diary of 1941 that the discord with the 25th Army was a matter of consequence, and thus he was willing to be an 'abuse receiver' since his primary goal was to accomplish his mission.[35] Sugawara's opinion indicates that he saw the Imperial Army firmly taking an old-fashioned view that could not see what the potential of air power really was.

Sugawara's leadership in the Malaya offensive air operation

Sugawara emphasised 'aerial exterminating action' as his concept of primary operations, but this concept ran counter to the influential Imperial Army philosophy at the time. This meant that Sugawara could not directly force the implementation of his ideas during the Malaya offensive air operation. Rather, his basic attitude of operational leadership was to give a free hand to his subordinate commanders to the maximum possible extent based on mutual understanding and reliance, and then to inspire a positive autonomous posture to achieve their mission. Writing in his diary, as Sugawara looked back on his own operational leadership from the beginning of the war to the end of 1941, he concluded and regretted that the results of the air operations were not satisfactory, especially as he compared it to the Philippine theatre.[36] He blamed himself for insufficient leadership. By 'insufficient leadership' he meant his regret that he was unable to visit each of his subordinate units after assuming his responsibility, despite his wish to build good relationships with his subordinate commanders.

Sugawara's leadership attached more importance to these relationships than to his operational concept of pre-emptive attack by 'aerial exterminating action'. On D-Day, Major-General Kenji Yamamoto, Commander of the 7th Air Brigade, led three heavy bomber air regiments and one fighter air regiment. Yamamoto directed a suspension of all the planned 4 a.m. take-offs, and changed the mission to a daylight strike owing to poor runway conditions caused by intensive heavy rain from the previous day. At this moment, Sugawara tried to persuade Yamamoto to let his (Yamamoto's) regiments with good runway conditions take off to concentrate Sugawara's Air Corps' strike power. However, Yamamoto did not agree with Sugawara's opinion. Therefore Sugawara considered directing part of the air regiments of the 7th Air Brigade to take off by issuing an order to put them directly under the command of the 3rd Air Corps. In the meantime, as it turned out, one heavy bomber air regiment of the 7th

Air Brigade had already taken off toward its target owing to a delay in relaying Yamamoto's directive. Sugawara abandoned his plan to issue a higher-level order, did not divide the 7th Air Brigade into three parts, and supported the decision by Yamamoto. Sugawara wrote in his diary that he had wanted to override Yamamoto's directive but he changed his mind. He felt that it was not good for future operations to take drastic measures.[37] On the other hand, he wrote in his diary at the end of 1941 that the results of the air operations were not satisfactory. He gave as one of the reasons that British air forces evaded air battles virtually from the beginning of the war. He stated later on that, if the enemy avoids hostilities, it might be possible to 'attack' but it is not so easy to 'exterminate' the enemy's air power.[38] For Sugawara it was really important to accomplish a pre-emptive raid at the beginning of the confrontation, but he allowed his priority to relationships to override his principles of operational concept.

Another characteristic of Sugawara's leadership was that he wanted his subordinate commanders to actively command independently. His typical attitude was shown to Major-General Saburo Endo, Commander of the 3rd Air Brigade. Against criticism from the Southern General Army Headquarters saying that Endo's aggressiveness brought unnecessary attrition of air assets, Sugawara wrote in his diary on New Year's Day in 1942 that he had a different opinion from Endo on air power build-up, but that he respected Endo's earnest posture of operational leadership, and never sought to blame him.[39] During this period, each airfield of the 3rd Air Corps was under frequent air raid attack by British forces, and, more than the other brigades, the 3rd Air Brigade suffered severe damage to more than 50 per cent of their aircraft.[40] Facing this situation, Sugawara directed his subordinate commanders on 28 December 1941 to keep going and to do their best to try to preserve the present force level. On 27 December, the previous day, Endo had presented his opinion on the capture of airfields by the Kuantan landing operation in order to carry out Singapore's 'aerial exterminating action' from the airfield. So, when he received General Sugawara's directive of 28 December, he ignored it while the 3rd Air Brigade continued a tough mission under enemy air raid in the vicinity of enemy airfields. Endo thought that pursuing the ongoing mission while trying to preserve present force levels represented a contradiction in objectives. Sugawara disagreed with Endo's opinion because of the difference in judgement about the entire situation.[41] However, since Sugawara put the 3rd Air Brigade as a forward deployment force, he tried to give Endo a more free hand, even though he had ordered subordinates to maintain current force levels to the maximum possible extent.[42] The reason why Sugawara was generous to Endo was because Endo's command concept and his concept on the use of air power were almost identical to Sugawara's operational concept and thus should have paved the way to undoubted victory by performing offensive raids.

Sugawara's posture regarding the ground support mission was to cope with requests from them flexibly without restraining air units all the time. However, on one obvious occasion, Sugawara showed his clear judgement on support to ground forces when the Kuantan and Mersing landing operations were planned again to capture Singapore. Along with the southbound offensive advance on the peninsula, the operation plan focused on a seaborne landing by almost one infantry division of the Southern General Army. The Imperial Army–Navy operational agreement was developed for this purpose. Sugawara forwarded his opinion to General Terauchi Hisaichi, Commander of Southern General Army, that he should suspend the additional landing operation. The reason for his opposition was the 3rd Air Corps status at that time. Sugawara thought there was no question about the necessity of the Imperial Army–Navy joint air strike over the Singapore area, but at that moment his air corps had just accomplished the Rangoon air raid and needed more time to regenerate air assets before another strike. Sugawara felt they were too poorly prepared – especially in ensuring there were enough airfields to perform 'aerial exterminating action' toward Singapore.[43] Meanwhile, the Kuantan landing operation itself was cancelled until Takumi Column could successfully advance to a point of approximately 10 kilometres north of Kuantan. Sugawara never intended to refuse the ground support mission, but he stressed 'aerial exterminating action' in his operational leadership. Even as he recommended suspension of the operation to Terauchi, he also forwarded his plan to participate in the operation with his reduced air power. Sugawara was confident he had the flexibility to shift to support of ground forces if favourable progress of the ground operation was not achieved by advance of the Takumi Column.

There were only two occasions during the Malaya offensive air operation when Sugawara made strong recommendations based on position authority: one was his suspension directive on departures by the 7th Air Brigade at the beginning of the war, and the other was expressing his opinion on cancellation of the Kuantan landing operation. Though he complained and wrote of some worries in his diary about the discord or feud within the 25th Army and his subordinate commanders, General Sugawara settled them himself as matters of his inner emotional condition. The basis of his operational leadership was mutual understanding and reliance with his commanders. The typical example was the Kuala Lumpur air raid on 22 December led by Lieutenant-Colonel Kato Takeo, Commander of 64th Air Regiment. Complete seizure of control of the air over Malaya was accomplished with this air strike.

The pitfalls in Tanigawa and Sugawara's operational concept of an 'air force'

The Imperial Army turned to defensive operations after it conquered Sumatra, Java and Burma. The operational concept to adopt 'aerial exterminating action' as the main tactic revealed a serious weak point in the air defence operation. Its weak point was the Imperial Army's traditional thought on making light of defensive operation. 'Air Troops Operation' as the prescription for leadership in 1937 strictly advocated a passive posture such as interception and pursuit after coming under attack from the enemy. The succeeding manual of 1940, 'Air Operation Essentials', mentioned air defence for vital military, political and economical areas, but insisted on timely 'air exterminating action' as the way to achieve this purpose.[44] This thought came from lessons of the Nomonhan incident as follows: the proportion of interception and air raid in military results was one-to-six; therefore the offensive was indispensable in gaining the initiative in an operation.[45]

The preference toward 'aerial exterminating action' led to a delay in the building up of air defence systems. The Imperial Army unexpectedly discovered a certain document in one of the anti-aircraft gun positions right after they captured Singapore. The document described the principle of a pulse-radar to direct a searchlight. The Imperial Army reproduced this document and delivered it to research agencies, headquarters and so on because of its technical impact. This document was called the 'Newman note'.[46] A radar system based on this principle was in the development phase for early warning in the Imperial Army at that time, but anti-aircraft fire radar was not. Immediately the Imperial Army started test-manufacturing an imitation of this radar and then completed it as a fixed anti-aircraft gunfire radar, the 'Tachi-2'. But the ability to develop this radar was limited to a level that only produced 20 sets in January 1943.[47] Sugawara had referred to the necessity of an air defence system with radars and anti-aircraft artillery in his report to the Commander of Southern General Army in May 1943.[48] The Imperial Army realised the importance of an organised air defence system in addition to 'aerial exterminating action' in revisions of 'Air Operation Essentials' in 1944.[49]

Although Imperial Army Aviation tried independent operations as an 'air force' in the Malaya offensive air operation, they unfortunately failed to fully recognise the importance of systematic air defence operations because of their overemphasis on 'air exterminating action', and the blinding euphoria over their brilliant victory with relatively great ease in Malaya.

Conclusion

The Malaya offensive air operation was a magnificent experiment in that Imperial Army Aviation broke through the framework of the Imperial Army to become an 'air force'. This experiment was one of independent air operations that gained control of the air by means of 'aerial exterminating action'. Independent air operations as an 'air force' were key to the success of the Imperial Army's operation in Malaya. However, Tanigawa and Sugawara, who led the experiment, never showed powerful leadership in the Malaya offensive air operation, even though they had great resolution and the principles of an operational concept to become an 'air force'. Although the environment of the Army was essentially different from that of an air force, they successfully led Imperial Army Aviation to independent air operations as an 'air force' by co-ordination with higher echelons, subordinate units and ground forces, and co-ordination between 'aerial exterminating action' and ground support missions. However, the result was that even Sugawara and Tanigawa could not break through the Imperial Army's hard shell of a traditional operational concept.

From the experience of air operations during the early stages of the Sino-Japanese War of 1937–45, Sugawara gained the insight that the 'soil' of the Imperial Army was not adequate to bring up an 'air force'. He stated his opinion to the Minister of War in 1939. He wrote in his report metaphorically:[50]

> An army is a 'gamecock', a navy is a 'hawk' and an air force is an 'eagle'. A 'hawk' can foster a young 'eagle' as its own, but a 'gamecock' does not know how to bring up an 'eagle'. A 'gamecock' tends to think of pulling off feathers to strengthen the legs; consequently a 'gamecock' can never bring up an 'eagle' as a real 'eagle'.

Then referring to the essential difference between an air force and ground forces, he condemned the Imperial Army's ground forces for misunderstanding the difference:

> The core of a ground forces army is manpower, but that of an air force is aircraft itself. If aircraft do not show the power, an air force will never be able to become a real air force. The air force cannot work only by manpower. The Imperial Army ground forces do not understand this essential difference at all. Therefore it will have conflict with our aviation. Their misunderstanding would have our aviation to 'attack by reconnaissance planes if we lost our bombers and fighters, even though reconnaissance planes have no gun'. Army insistence derives from their narrow view based on the Imperial Army's motto on education and training, 'Charge your enemy when your ammunition runs out.'

Sugawara saw the last day of the Pacific War as Commander of the 6th Air Army. This air army was the special air strike corps (Tokkohtai) in the Okinawan theatre. When we remember that he condemned the Imperial Army's insistence on this as a suicide attack, we can see here the irony of history.

Appendix 10.1

Michioho SUGAWARA – a chronology

1888, Nov. 28	Born in Nagasaki Prefecture, Kyushu
1909, May	Graduated from Military Academy
1909, Dec.	Commissioned; attached to 4th Infantry Regiment
1913, Feb.	Lieutenant (Infantry)
1916–19	Military Staff College
1919, June	Captain; Company Commander, 4th Infantry Regiment
1924, Jan.–May	To the United States
1924, Aug.	Major (Infantry)
1925, March	Battalion Commander, 76th Infantry Regiment
1925, May	Transferred to air service; Major (Aviation); attached to 6th Air Regiment
1925, Dec.	Officer-student in the higher command course at Military Staff College
1927, Dec.	Member of Army Air Service HO
1928, Aug.	Lieutenant-Colonel (aviation)
1931, Aug.	Instructor, Shimoshizu Army Flying School
1933, March	Colonel (aviation); Captain, 6th Group
1935, March	Chief, 1st Section, Army Air Service HO
1936–37	To Europe; Assistant Head, Oshima Air System Inspection Delegation to Germany
1937, Aug.	Major-General; Commander, 2nd Air Brigade
1938, July	Commander, 3rd Air Brigade
1939, Oct.	Lieutenant-General; attached to Head office of the Inspector-General of Military Air Training
1939, Dec.	Commandant, Shimoshizu Army Flying School
1940, Aug.	Commander, 1st Air Corps
1941, Sep.	Commander, 3rd Air Corps
1942, April	Commander, 3rd Air Division
1942, July	Commander, 3rd Air Army
1943, May	Commandant, Military Air Academy
1944, March	Deputy-Chief of the Army Air Service HO
1944, July	Concurrently Inspector-General of Military Air Training
1944, Aug.	Concurrently Commander, Training Air Army
1944, Dec.	Commander, 6th Air Army
1945, Dec.	Demobilized
1983, Dec. 29	Died, aged 95

NOTES

1 BRITAIN'S GRAND STRATEGY AND ANGLO-AMERICAN LEADERSHIP IN THE WAR AGAINST JAPAN

1 Christopher Thorne, *The Far Eastern War: States and Societies 1941–1945* (London: Unwin, 1986), pp. 4–10.

2 John Keegan, 'Churchill's Strategy', in Robert Blake and Wm. Roger Louis (eds), *Churchill: A Major New Assessment of his Life in Peace and War* (Oxford: Oxford University Press, 1993), p. 337.

3 For a brief review on the Pacific War, see Saki Dockrill, 'One Step Forward: A Reappraisal of the "Pacific War"'. in Saki Dockrill (ed.), *From Pearl Harbor to Hiroshima: The Second World War in Asia and the Pacific 1941–1945* (Basingstoke: Macmillan, 1994), pp. 1–7 ff. For more pessimistic views on the British Empire, etc., see Dick Wilson, 'Churchill Belittled Threat of Pacific War', *Japan Times*, 27 August 1992, and also Correlli Barnett, *The Collapse of British Power* (Gloucester, UK: Alan Sutton, 1972). While it is true that the increasing number of recent studies take international perspectives on the Pacific War, including those of the British, they tend to focus on the causes, and consequences, of the 1941–45 war, rather than the conduct of the war itself. For instance, C. Hosoya, A. Iriye, N. Honma *et al.* (eds), *Taiheiyo Senso* [The Pacific War] (Tokyo: Tokyo University Press, 1993). See also a three-volume history on the Pacific War (the causes, the outbreak and the end of the war) edited by the Military History Society of Japan in 1990. Two books stand out in the literature of the Pacific War, Christopher Thorne, *Allies of a Kind: The United States, Britain and the War against Japan, 1941–1945* (London: Hamish Hamilton, 1978), and Ronald H. Spector, *Eagle against the Sun: The American War with Japan* (Middlesex: Penguin, 1987), which, while focusing on the US conduct of war, also deals with the roles of Britain and China. For a general account of the Pacific War, see Guy Wint and John Pritchard, *Total War: The Causes and Courses of the Second World War: The Greater East Asia and Pacific Conflict* (vol. 2) (London: Penguin, 1989).

4 Paul Kennedy, 'Grand Strategy in War and Peace: Towards a Broader Definition', in Paul Kennedy (ed.), *Grand Strategies in War and Peace* (New Haven, CT: Yale University Press, 1991), pp. 1–5 ff.

5 For a concise account of Britain's way in warfare, see David French, *The British Way in Warfare, 1688–2000* (London: Unwin Hyman, 1990), pp. xi–xviii.

6 Bernard Porter, *The Lion's Share: A Short History of British Imperialism, 1850–1983* (London: Longman, 1984), pp. 1–7; John Darwin, *Britain and Decolonisation: The Retreat from Empire in the Post-War World* (Basingstoke: Macmillan, 1988), p. 25.

7 French, *The British Way in Warfare*, p. 175; Paul Kennedy, *The Rise and Fall of the Great Powers* (New York: Random House, 1987), pp. 275–91.

8 French, *The British Way in Warfare*, pp. 179–80, 185–6; David Reynolds, 'Britain, the Pacific, and Appeasement', in Hosoya *et al.* (eds), *Taiheiyo Senso*, pp. 312–3.

9 James Neidpath, *The Singapore Naval Base* (Oxford: Clarendon, 1981), pp. 28–31; David Dilks (ed.), *Retreat from Power, 1906–1939* (London: Macmillan, 1981), pp. 12–14.

10 Winston S. Churchill, *The Second World War: The Grand Alliance* (vol. III) (London: Cassell, 1950), p. 516.

11 Anthony Best, *British Intelligence and the Japanese Challenge in Asia, 1914–1941* (Basingstoke: Macmillan/Palgrave, 2002), pp. 83–104.

12 'Defence Requirements Sub-Committee Report', 28 Feb. 1934, DRC 14, CAB 16/109, cited in Michael Howard, 'British Military Preparations for the Second World War', in Dilks (ed.), *Retreat from Power*, pp. 108–9.

13 French, *The British Way in Warfare*, p. 194; Shigeru Hayashi, *Taiheiyo Senso* [The Pacific War] (Tokyo: Chuko bunko, 1980), pp. 173–83; Akira Fujiwara, *Nihon Gunj Shi* [Japanese Military History] (vol. I) (Tokyo: Nihon Hyoron sha, 1987), pp. 254–67.

14 Martin Gilbert, *Finest Hour: Winston S. Churchill, 1939–1941* (London: Heinemann, 1983), p. 1118.

15 WP (39)148, 28 Nov. 1939, CAB 80/5, cited in John Pritchard, 'Winston Churchill, the Military and Imperial Defence in East Asia', in Dockrill (ed.), *From Pearl Harbor to Hiroshima*, p. 34.

16 Pritchard, 'Winston Churchill, the Military and Imperial Defence', pp. 41–2.

17 Peter Lowe, *Great Britain and the Origins of the Pacific War* (Oxford: Clarendon, 1977), pp. 6–7.

18 Ian Nish, 'British Thoughts on the Ending of the Asia-Pacific War', in *Dainiji Sekaitaisen-Shusen* [The End of the Second World War] (vol. III), ed. Gunji Shigaku Kai [The Military History Society of Japan] (Tokyo: Kinsei sha, 1990), p. 282.

19 Pritchard, 'Winston Churchill, the Military and Imperial Defence', p. 43.

20 Saki Dockrill, 'Hirohito, the Emperor's Army and Pearl Harbor', *Review of International Studies*, 18 (1992), pp. 324–8.

21 Boeicho-Senshi shsitsu [Military Historical Branch, Japanese Self-Defense Agency), *Hawaii Sakusen* [Strategy for Hawaii] (Tokyo: Asagumo Shinbunsha, 1967), pp. 12–19, 189–93; Beoeicho sensshi bu (ed.) [Military Historical Branch, the Japanese Self-Defense Agency], *Daihon'ei Kaigunbu-Rengo Kantai* [The Imperial Navy General Headqurters – the Combined Fleet] (2) (Tokyo: Asagumo shinbushan, 1985), p. 523. See also Dockrill, 'Hirohito, the Emperor's Army and Pearl Harbor', pp. 329–32.

22 For a recent study on Churchill's leadership, see Eliot A. Cohen, *Supreme Command: Soldiers, Statesmen and Leadership in Wartime* (New York: Free Press, 2002), pp. 95–132.

23 Maurice Matloff, 'Allied Strategy in Europe, 1939–1945', in Peter Paret (ed.), *Makers of Modern Strategy: From Machiavelli to the Nuclear Age* (Princeton, NJ: Princeton University Press, 1986), pp. 691, 697.

24 Spector, *Eagle against the Sun*, p. 127; R. A. C. Parker, *Struggle for Survival: The History of the Second World War* (Oxford: Oxford University Press, 1989), p. 117; John Charmley, *Churchill: The End of Glory: A Political Biography* (London: Hodder & Stoughton, 1993), pp. 475–6.

25 Warren F. Kimball, 'Wheel within a Wheel: Churchill, Roosevelt, and the Special Relationship', in Blake and Louis (eds), *Churchill*, pp. 296–9; Matloff, 'Allied Strategy', p. 680; D. Clayton James, 'American and Japanese Strategies in the Pacific War', in Paret (ed.), *Makers of Modern Strategy*, p. 711.

26 Parker, *Struggle for Survival*, p. 122; Churchill, *The Second World War* (vol. III), p. 622.

27 Churchill, *The Second World War* (vol. III), p. 515.

28 Kimball, 'The Special Relationship', p. 299.

29 Churchill, *The Second World War* (vol. III), pp. 622, 579.

30 Eliot A. Cohen, 'Churchill and Coalition Strategy in World War II', in Kennedy (ed.), *Grand Strategies*, p. 50.

31 Matloff, 'Allied Strategy', p. 682.

32 Martin Gilbert, *Road to Victory: Winston S. Churchill, 1941–1945* (London: Heinemann, 1986), pp. 31–2.

33 Spector, *Eagle against the Sun*, p. 327; Wenzhao Tao, 'The China Theatre and the Pacific War', in Dockrill (ed.), *From Pearl Harbor to Hiroshima*, p. 134.

34 Clayton James, 'Strategies in the Pacific War', p. 720; Spector, *Eagle against the Sun*, pp. 328–30 ff.

35 French, *The British Way in Warfare*, pp. 198, 205.

36 Gilbert, *Road to Victory*, pp. 29, 41, 46–7. See also Robert O'Neill, 'Churchill, Japan, and British Security in the Pacific, 1904–1942', in Blake and Louis (eds), *Churchill*, pp. 275–89.

37 Gilbert, *Road to Victory*, p. 61; Y. Kojima, *Taiheiyo senso* (vol. I) (Tokyo: Chuko shinsho, 1988), pp. 158–9; Wint & Prichard, *Total War* (vol. II), pp. 394–7 ff; Allen Louis, *Singapore 1941–1942* (London: Davis-Poynter, 1977), pp. 15–22 ff.

38 Gilbert, *Road to Victory*, p. 61.

39 Clayton James, 'Strategies in the Pacific', p. 723; Spector, *Eagle against the Sun*, p. 143.

40 *Spector, Eagle against the Sun*, p. 142.

41 Brian Bond, *The Pursuit of Victory* (Oxford: Oxford University Press, 1996), p. 160; Matloff, 'Allied Strategy', p. 679.

42 Parker, *Struggle for Survival*, pp. 121–3; Kimball, 'The Special Relationship', p. 301.

43 Gilbert, *Road to Victory*, pp. 293–4, 308; Parker, *Struggle for Survival*, p. 124.

44 Gilbert, *Road to Victory*, pp. 307–8.

45 Parker, *Struggle for Survival*, p. 123.

46 Spector, *Eagle against the Sun*, p. 349; Thorne, *Allies of a Kind*, p. 227.

47 Wenzhao Tao, 'The China Theatre', in Dockrill (ed.), *From Pearl Harbor to Hiroshima*, p. 135; Spector, *Eagle against the Sun*, p. 327.

48 Thorne, *Allies of a Kind*, pp. 225–6; Wenzhao Tao, 'The China Theatre', pp. 135–6.

49 Spector, *Eagle against the Sun*, pp. 342–3.

50 Thorne, *Allies of a Kind*, p. 226.

51 Jonathan Spence, *The Search for Modern China* (London: Hutchinson, 1990), p. 471.

52 Wesley Bagby, *The Eagle–Dragon Alliance: America's Relations with China in World War II* (Newark, DE: University of Delaware, 1992), p. 58.

53 Spector, *Eagle against the Sun*, p. 349; Bagby, *The Eagle–Dragon Alliance*, pp. 71–3; Wenzhao Tao, 'The China Theatre', p. 139.

54 Gilbert, *Road to Victory*, pp. 400–403; Bagby, *The Eagle–Dragon Alliance*, pp. 73–77.

55 Spector, *Eagle against the Sun*, pp. 336–8; Parker, *Struggle for Survival*, p. 122.

56 Philip Ziegler (ed.), *Personal Diary of Admiral the Lord Louis Mountbatten: Supreme Allied Commander, South-East Asia, 1943–1946* (London: Collins, 1988), p. xiii.

57 Spector, *Eagle against the Sun*, p. 352.

58 For the Cairo conference, see Keith Sainsbury, *The Turning Point: Roosevelt, Stalin, Churchill and Chiang Kai-Shek, 1943, The Moscow, Cairo, and Teheran Conferences* (Oxford: Oxford University Press, 1985). As for Mountbatten's offer, see Gilbert, *Road to Victory*, p. 595; Philip Ziegler, *Mountbatten* (London: Fontana, 1985), pp. 261–3.

59 Sainsbury, *The Turning Point*, p. 206; Parker, *Struggle for Survival*, p. 187.

60 Sainsbury, *The Turning Point*, p. 246; Gilbert, *Road to Victory*, pp. 599–600; Bagby, *The Eagle–Dragon Alliance*, p. 91.

61 Winston S. Churchill, *Second World War: Triumph and Tragedy* (vol. VI) (London: Cassell, 1954), pp. 135–7.

62 Wint and Pritchard, *Total War* (vol. II), pp. 569–70.

63 Churchill, *Triumph and Tragedy*, p. 543.

64 Ibid., pp. 340–2; A. W. Purdue, *The Second World War* (Basingstoke: Macmillan, 1999), p. 169.

65 Churchill, *Triumph and Tragedy*, pp. 552–3.

66 Ibid., p. 130.

67 Lawrence Freedman, 'The Third World War?', *Survival*, 43, 4 (Winter 2001), pp. 70–4.

68 For the British plan to withdraw from Singapore, and Allied reactions, see Saki Dockrill, *Britain's Retreat from East of Suez: The Choice between Europe and the World?* (Basingstoke: Macmillan/Palgrave, 2002).

2 TOJO HIDEKI AS A WAR LEADER

1 A. J. P. Taylor, *The War Lords* (London: Hamish Hamilton, 1977), p. 158.

2 David A. Titus, *Palace and Politics in Prewar Japan* (New York: Columbia University Press, 1974), p. 11.

3 Shiobara Tokisaburo, *Tojo Memo* [Tojo's Memoir], cited in Tojo Hideki Kanko-kai and Joho Kaio (eds), *Tojo Hideki* (Tokyo: Fuyo-shobo, 1974), p. 341.

4 Amemiya Shoichi, *Kindai Nihon no Senso Shido* [War Leadership in Modern Japan] (Tokyo: Yoshikawa-kobunkan, 1997), ch. 1.

5 Sato Kenryo, *Daitoa-senso Kaiko-roku* [The Greater East Asia War Memoirs] (Tokyo: Tokuma-shoten, 1966), p. 179; Sato Kenryo, *Sato Kenryo no Shogen* [Testimony of Sato Kenryo] (Tokyo: Fuyo-shobo, 1976), pp. 366–7.

6 See Ito Takashi *et al.*, (eds), *Tojo Naikaku Sori-daijin Kimitsu Kiroku/Tojo Hideki Taisho Genko-roku* [Prime Minister Tojo's Secret Documents and Chronicle of General Tojo's Sayings and Doings] (Tokyo: Tokyo-daigaku-shuppankai, 1990).

7 Sato, *Daitoa-senso Kaiko-roku*, p. 264. A few months before, Tojo had a bitter argument with Nagano Osami, the Chief of the Navy General Staff, on the occupation of Portuguese Timor. Nagano insisted on occupying the island because of operational necessity. But Tojo objected to it since it might drive neutral Portugal into the enemy camp. Tojo and Nagano did not talk to each other for a while after this dispute. Tojo might have been convinced by this dispute that it was most important to avoid conflicts between two services at the top level. Tanemura Sako, *Daihon'ei Kimitsu Nisshi* [Secret Diary of the IGHQ] (Tokyo: Daiyamondo-sha, 1952), pp. 115–16.

8 Boei-cho Boeikenshu-sho Senshi-shitu [War History Office, National Defense College, Defense Agency, Japan], *Senshi-sosho: Daihon'ei Rikugun-bu* [Series of Military History of World War II: The Army Department of the IGHQ] (vol. V) (Tokyo: Asagumo-shinbun-sha, 1973), p. 309; Hayashi Saburo, *Taiheiyo-senso Rikusen Gaishi* [General History of Land Campaigns in the Pacific War] (Tokyo: Iwanami-shoten, 1951), pp. 101–2.

9 Nishiura Susumu, *Showa Senso-shi no Shogen* [Witness of War History in Showa Era] (Tokyo: Hara-shobo, 1980), pp. 193–4.

10 Imoto Kumao, *Sakusen Nisshi de tsuzuru Daitoa-senso* [The Greater East Asia War Told by Operation Diary] (Tokyo: Fuyo-shobo, 1979), p. 517.

11 Tanemura, *Daihon'ei Kimitsu Nisshi*, p. 169.

12 'Tai-Bei-Ei-Ran-Sho Senso Shumatsu Sokushin ni kansuru Fukuan' [Plan to Hasten the End of the War with the United States, Britain, Holland and China; The Liaison Conference Decision on 15 November 1941], in Sanbohonbu [Army General Staff] (ed.), *Sugiyama Memo: Daihon'ei Seifu Renrakukaigi-to Hikki* [General Sugiyama's Memoir: Minutes of Liaison Conferences and Other Meetings] (vol. I) (Tokyo: Hara-shobo, 1967), pp. 523–5.

13 Sato Sanae, *Tojo Hideki: Huin sareta Shinjitu* [Tojo Hideki: His Sealed Truth] (Tokyo: Kodan-sha, 1995), p. 76.

14 *Sugiyama Memo* (vol. II), p. 16.

15 Ibid., p. 51.

16 *Senshi-sosho: Daihon'ei Rikugun-bu*, (vol. III) (1970), p. 342.

17 'Kongo toru-beki Senso-shido Hoshin no Taiko' [General Principles of the War Strategy to be Adopted in the Next Stage; The Liaison Conference Decision on 7 May 1942], in *Sugiyama Memo*, (vol. II), pp. 81–2.

18 *Senshi-sosho: Daihon'ei Rikugun-bu* (vol. IV) (1972), p. 356.

19 Sato Sanae, *Tojo Hideki*, pp. 72–4.

20 *Senshi-sosho: Daihon'ei Rikugun-bu* (vol. IV), p. 406.

21 *Sugiyama Memo* (vol. II), p. 379.

22 'Kongo toru-beki Senso-shido Hoshin no Taiko' [General Principles of the War Strategy to be Adopted in the Next Stage; The Imperial Conference Decision on 30 September 1943]] *Sugiyama Memo* (vol. II), p. 473.

23 Ibid., pp. 465, 467.

24 *Senshi-sosho: Daihon'ei Rikugun-bu* (vol. III), p. 552.

25 Goto Ken'ichi, 'Tojo Shusho to "Nanpo Kyoei-ken"' [Prime Minister Tojo and 'South Co-Prosperity Sphere'], in Peter Duus and Kobayashi Hideo (eds), *Teikoku to iu Genso: 'Daitoa Kyoei-ken' no Shiso to Genjitsu* [An Illusion of an Empire: The Ideas and Realities of Greater East Asia Co-Prosperity Sphere], (Tokyo: Aoki-shoten, 1998), p. 278.

26 *Sugiyama Memo* (vol. II), p. 515.

27 Goto, 'Tojo Shusho', p. 273.

28 *Senshi-sosho: Daihon'ei Rikugun-bu*, (vol. VI) (1973), p. 109.

29 For some detail of the operation, see Tobe Ryoichi *et al.*, *Shippai no Honshitsu: Nihon-gun no Soshikiron-teki Kenkyu* [*Essence of Failure: Organisational Approach to the Japanese Armed Forces*] (Tokyo: Daiyamondo-sha, 1974), pp. 92–119.

30 *Senshi-sosho: Daihon'ei Rikugun-bu* (vol. VII) (1973), p. 101.

31 *Senshi-sosho: Inparu Sakusen* [Imphal Campaign] (1968), p. 159.

32 *Senshi-sosho: Daihon'ei Rikugun-bu* (vol. VIII) (1974), p. 245.

33 Ibid., p. 301.

34 *Senshi-sosho: Inparu Sakusen*, pp. 520–1; Tanemura, *Daihon'ei Kimitsu Nisshi*, pp. 168–9.

35 *Senshi-sosho: Inparu Sakusen*, p. 521.

36 *Sato Kenryo no Shogen*, pp. 399–400.

37 As for Tojo's loyalty to the Emperor, see Hatano Sumio, 'Seiji Shido-sha toshite no Tojo Hideki' [Tojo Hideki as a Political Leader], *Meiji Seitoku Kinen Gakkai Kiyo* [Journal of Society for Commemoration of the Emperor Meiji's Virtues], 35 (June 2002).

38 Nishiura, *Showa Senso-shi no Shogen*, pp. 147, 171, 145.

39 Akamatu Sadao, *Tojo Hisho-kan Kimitu Nisshi* [Secret Diary of Tojo's Secretary] (Tokyo: Bungei-Shunju, 1985), p. 37.

40 Ito Takashi, *Tojo Naikaku Sori-daijin Kimitsu Kiroku/Tojo Hideki Taisho Genko-roku*, p. 19.

41 Ibid., pp. 509, 519.

3 THE ARMY LEVEL OF COMMAND: GENERAL SIR WILLIAM SLIM AND FOURTEENTH ARMY IN BURMA

1 G. D. Sheffield (ed.), *Leadership and Command: The Anglo-American Military Experience since 1861* (London: Brassey's, 1997), pp. 10–11, 126.

2 Sir John Smyth, VC, *Leadership in War, 1939–1945* (Newton Abbot: David & Charles, 1974), p. 221.

3 John Keegan (ed.) *Churchill's Generals* (London: Weidenfeld & Nicolson, 1991), pp. 298–322.

4 Basil Liddell Hart, *History of the Second World War* (London: Cassell, 1970).

5 Duncan Anderson, 'Slim', in Keegan, *Churchill's Generals*, pp. 298–300.

6 Ronald Lewin, *Slim: the Standardbearer* (London: Leo Cooper, 1976), p. 190.

7 Ibid., pp. 45, 50, 194.

8 Anderson, 'Slim', pp. 301–2.

9 Anderson, 'Slim', pp. 303–4. Lewin, *Slim*, pp. 53–4.

10 Lewin, *Slim*, pp. 64–7.

11 Anderson, 'Slim', pp. 304–5. Geoffrey Evans, *Slim as Military Commander* (London: Batsford, 1969), pp. 38–41.

12 Evans, *Slim as Military Commander*, pp. 51–2.

13 Anderson, 'Slim', pp. 307–10.

14 Field Marshal Sir William Slim, *Defeat into Victory* (London: Cassell, 1956), pp. 102–10.

15 Ibid., p. 86.

16 Lewin, *Slim*, p. 90.

17 Evans, *Slim as Military Commander*, pp. 82–4.

18 Slim, *Defeat into Victory*, pp. 115–21.

19 Ibid., p. 146.

20 Ibid., pp. 182–7.

21 Anderson, 'Slim', pp. 312–14.

22 General Sir Rupert Smith, 'Should Generals be "Slim" Today?' Lecture at the Imperial War Museum, 9 February 2002.

23 Slim, *Defeat into Victory*, pp. 207–8.

24 Ibid., p. 213.

25 Lewin, *Slim*, pp. 128–9. Evans, *Slim as Military Commander*, p. 105.

26 Slim, *Defeat into Victory*, pp. 216–20. Lewin, *Slim*, pp. 142–4.

27 See, for example, David Rooney, 'Command and Leadership in the Chindit Campaigns', in Sheffield, *Leadership and Command*, pp. 141–57.

28 Anderson, 'Slim', p. 314.

29 Slim, *Defeat into Victory*, pp. 294–308.

30 Evans, *Slim as Military Commander*, pp. 155–7.

31 Slim, *Defeat into Victory*, p. 310. Sato may have been too doggedly fixated on Kohima but he also showed moral courage in refusing to divert units to Imphal, and eventually forfeited his command by insisting that the remnants of his division must retreat. See Lewin, *Slim*, pp. 185–7.

32 Lewin, *Slim*, p. 176.

33 Slim, *Defeat into Victory*, pp. 366–9. Lewin, *Slim*, p. 184. Evans, *Slim as Military Commander*, pp. 176–7.

34 Lewin, *Slim*, pp. 198–201.

35 Lewin, *Slim*, pp. 208–9. Evans, *Slim as Military Commander*, pp. 184 ff. Slim, *Defeat into Victory*, pp. 390–3.

36 Lewin, *Slim*, pp. 210–13.

37 Evans, *Slim as Military Commander*, pp. 187–8, 214–15. Anderson, 'Slim', pp. 318–19.

38 Slim, *Defeat into Victory*, pp. 428–9. The first attempt to cross the Irrawaddy failed completely but fortunately the Japanese had already abandoned the city of Pagan on the opposite bank. This was the longest river crossing in the Second World War.

39 Lewin, *Slim*, pp. 237–46. Rupert Smith, in the lecture mentioned earlier, suggests that although Slim was not a 'dabbler in politics' he *was* politically astute. It was a remarkable feat to be 'sacked' first as a corps and then as an Army commander, on both occasions to take the position of the general who had tried to dismiss him.

40 See Michael Howard's comments in Sheffield, *Leadership and Command*, pp. 120, 126–7.

41 Mountbatten had no doubt that Slim was the finest. See Evans, *Slim as Military Commander*, pp. 215–15.

4 LEADERSHIP IN JAPAN'S PLANNING FOR WAR AGAINST BRITAIN

1 Sanbo Honbu [General Staff] (ed.), *Sugiyama Memo* [Army Chiefs of Staff General Sugiyama's memorandum and related documents on the Liaison Conference] (vol. I) (Tokyo: Hara Shobo, 1967), p. 315; Boeicho Boeikenshusho Senshishitsu *Senshi-sosho: Daihon'ei Rikugunbu* [Series of Military History of World War II: Imperial Headquarters, Army Department] (2) (Tokyo: Asagumo Shimbunsha, 1968), p. 433.

2 Major works on the Anglo-Japanese dimensions of World War II from Japanese perspectives are as follows. Kiyoshi Ikeda, 'Japanese Strategy and the Pacific War, 1941–5', Minoru Nomura, 'Military Policy-Makers Behind Japanese Strategy against Britain', both in Ian Nish (ed.), *Anglo-Japanese Alienation, 1919–1952* (Cambridge: Cambridge University Press, 1982), pp. 125–46, 147–55; Kiyoshi Ikeda, 'Anglo-Japanese Relations, 1941–45', in Ian Nish and Yoichi Kibata (eds), *The History of Anglo-Japanese Relations, 1600–2000*, vol. II: *The Political-Diplomatic Dimension, 1931–2000* (London: Macmillan, 2000), pp. 112–34; Hatano Sumio, 'Tai Ei Senso to "Dokuritsu Kosaku"', in Hirama Yoichi, Ian Gow and Hatano Sumio (eds), *Nichi Ei Koryu Shi*, vol. III: *Gunji* (Tokyo: Tokyo Daigaku Shuppan Kai, 2001), pp. 230–51.

3 Hata Ikuhiko, 'Senso Shumatsu Koso no Saikento', *Gunji-Shigaku* 31, 1–2 (September 1995), p. 22.

4 Sato Motohide and Kurosawa Fumitaka (eds), *GHQ Rekishika Chinjutsu Roku* (vol. II) (Tokyo: Harashobo, 2001), p. 861.

5 *Daihon'ei Rikugunbu* (2), p. 564, pp. 572–3; Sato and Kurosawa, *GHQ Rekishika Chinjutsu Roku* (vol. II), pp. 794–5.

6 *Sugiyama Memo* (vol. I), p. 523.

7 Ibid., pp. 523–24.

8 Hatano, 'Tai Ei Senso', p. 231; Ohki Takeshi, 'Doku-So Wahei Kosaku o meguru Gunzo' [Japanese Efforts to Mediate between Germany and the Soviet Union in 1942], *Nenpo Kindai Nihon Kenkyu*, 17 (1995), pp. 249–82.

9 *Daihon'ei Rikugunbu* (2), pp. 601–7.

10 Boeikenshusho Senshishitsu, *Senshi-sosho: Daihon'ei Kaigunbu, Rengo Kantai* [Imperial Headquarters, Navy Department, The Combined Fleet] (2) (Tokyo: Asagumo Shimbunsha, 1975), pp. 78–91; Hata Ikuhiko, 'Admiral Yamamoto's Surprise Attack and Japanese Navy's War Strategy', in Saki Dockrill (ed.), *From Pearl Harbor to Hiroshima: The Second World War in Asia and the Pacific, 1941–45* (London: Macmillan, 1994), pp. 62–4.

11 *Sugiyama Memo* (vol. I), p. 524.

12 Boeikenshusho Senshishitsu, *Senshi-sosho: Daihon'ei Rikubunbu* (3) (Tokyo: Asagumo Shimbunsha, 1970), p. 517.

13 Nishiura Susumu, *Showa Sensho Shi no Shogen* (Tokyo: Hara Shobo, 1980), p. 167.

14 Paul Kennedy, *Strategy and Diplomacy, 1870–1945* (London: George Allen & Unwin, 1983), p. 186.

15 Nomura Minoru, 'Taiheiyo Senso no Nihon no Senso Shido', *Nenpo Kindai Nihon kenkyu*, 4 (1982), pp. 37–8; Idem, *Nihon Kaigun no Rekishi* (Tokyo: Yoshikawa Kobun Kan, 2002), pp. 195–6.

16 *Daihon'ei Kaigunbu, Rengo Kantai* (2), pp. 333–5.

17 Ibid., pp. 307–24.
18 Ibid., p. 334; Ugaki Matome, *Sensho Roku* (Tokyo: Harashobo, 1968), pp. 58–91. Diary entry 30 December 1941; 5, 14, 27, 28 January; 22 February; 3 March 1942.
19 *Daihon'ei Kaigunbu, Rengo Kantai* (2), pp. 367–8; Hata, 'Admiral Yamamoto', p. 66.
20 Boeikenshusho Senshishitsu, *Senshi-sosho: Daihon'ei Rikugunbu* (3) (Tokyo: Asagumo Shimbunsha, 1970), pp. 518–19.
21 'Tanaka Shin'ichi Sanbohonbu Daiichi Bucho Gyomu Nisshi', diary entry 23 December 1941, Library of the National Institute for Defense Studies, Japan; *Daihon'ei Rikugunbu* (3), p. 33.
22 Hata, 'Senso Shumatsu Koso no Saikento', p. 28.
23 Boeikenshusho Senshishitsu, *Senshi-sosho: Ran'in Bengaru Wan Homen Kaigun Shinko Sakusen* [Assault from the Sea on the Dutch East Indies and the Bay of Bengal] (Tokyo: Asagumo Shimbunsha, 1969), ch. 9; Idem, *Senshi-sosho: Nansei Homen Kaigunsakusen* [Naval Operations in the Southwestern Theatre] (Tokyo: Asagumo Shimbunsha, 1976), pp. 641–78; Hans-Joachim Krug, Yoichi Hirama, Berthold J. Sander-Nagashima and Axel Niesté, *Reluctant Allies: German–Japanese Naval Relations in World War II* (Annapolis, MD: Naval Institute Press, 2001), pp. 43–57.
24 Nomura, *Nihon Kaigun no Rekishi*, pp. 196–7.
25 Boeikenshusho Senshishitsu, *Senshi-sosho: Daihon'ei Rikubunbu* (4) (Tokyo: Asagumo Shimbunsha, 1972), p. 281.
26 Boeikenshusho Senshishitsu, *Senshi-sosho: Daihon'ei Kaigunbu, Rengo Kantai* (3) (Tokyo: Asagumo Shimbunsha, 1974), pp. 23–6.
27 Ibid., pp. 55–60.
28 Ibid., pp. 141–2; Ugaki, *Senso Roku*, p. 160. Diary entry 7 August 1942.
29 *Sugiyama Memo* (vol. II), pp. 379–80; Boeikenshusho Senshishitsu, *Senshi-sosho: Daihon'ei Rikubunbu* (6) (Tokyo: Asagumo Shimbunsha, 1973), pp. 239–41.
30 Boeikenshusho Senshishitsu, *Senshi-sosho: Daihon'ei Rikubunbu* (6), p. 492.
31 Tomioka Sadatoshi, *Kaisen to Shusen* (Tokyo: Mainichi Shimbunsha, 1968), p. 56.
32 Sanbo Honbu, *Haisen no Kiroku* (Tokyo: Hara Shobo, 1967), p. 294.
33 Kurono Taeru, 'Daitoa Senso Kaisen Mae ni okeru Senso Shido Koso' [Japan's 'Strategy' for Fighting the Greater East Asia War: The Sequel to the Imperial Defence Policy], *Boei Kenkyujo Kiyo* [NIDS Security Studies], 2, 2 (September 1999), pp. 97–116.

5 CRISIS OF COMMAND: MAJOR-GENERAL GORDON BENNETT AND BRITISH MILITARY EFFECTIVENESS IN THE MALAYAN CAMPAIGN, 1941–42

1 Trevor Royle (ed.), *A Dictionary of Military Quotations* (New York: Simon & Schuster, 1989), p. 23.
2 A. P. Wavell, *Soldiers and Soldiering* (London: Jonathan Cape, 1953), p. 30.
3 S. Woodburn Kirby, *The War against Japan*, Vol. I: *The Loss of Singapore* (London: Her Majesty's Stationery Office, 1957) is the authoritative source. Recent accounts are Brian P. Farrell's ch. 8 in Malcolm H. Murfett *et al.*, *Between Two Oceans: A Military History of Singapore* (Singapore: Oxford University Press, 1999), and Alan Warren, *Singapore 1942: Britain's Greatest Defeat* (London: Hambledon, 2002). The

figures are from the 'Malayan campaign' entry in I. C. B. Dear (ed.), *The Oxford Companion to the Second World War* (Oxford: Oxford University Press, 1995).

4 For a succinct geopolitical analysis see my 'The Malayan Campaign, 1941–2, in International Perspective', *South Asia*, XIX, special issue (1996), pp. 169–82.

5 H. Gordon Bennett, 'Introduction', in Masonobu Tsuji, *Singapore: The Japanese Version* (London, Mayflower-Dell, 1966), p. vii. In fact, Yamashita's forces were just as likely to be in lorries as on bicycles.

6 Eventually published as H. Gordon Bennett, *Why Singapore Fell* (Sydney: Angus & Robertson, 1944).

7 Anon., cited in Wavell, *Soldiers and Soldiering*, p. 32.

8 The best biography is A. B. Lodge, *The Fall of General Gordon Bennett* (Sydney: Allen & Unwin, 1986).

9 See Keith Simpson, 'Percival', in John Keegan (ed.), *Churchill's Generals* (London: Warner Books, 1991), and Clifford Kinvig, *Scapegoat: General Percival of Singapore* (London: Brassey's, 1996).

10 Cited in Simpson, 'Percival', pp. 267, 269.

11 Bennett, *Why Singapore Fell*, pp. 19–20.

12 Ibid., p. 21.

13 Cited in Lodge, *Fall of General Gordon Bennett*, pp. 48–9.

14 Bennett, *Why Singapore Fell*, ch. v, esp. p. 22, also pp. 50, 75 and 82, and quotations cited in Lodge, *Fall of General Gordon Bennett*, pp. 74 and 78. Bennett was particularly critical of Heath's and Percival's decision to give up Kuala Lumpur, the Malayan capital, without a fight, and similarly he was later to criticise the giving up of the airfield at Kuantan.

15 Cited in Simpson, 'Percival', p. 266.

16 Bennett, *Why Singapore Fell*, p. 126.

17 Ibid., pp. 21, 23, 46, 157, 223.

18 Cited in Lodge, *Fall of General Gordon Bennett*, pp. 110 and 119.

19 Bennett, *Why Singapore Fell*, pp. 146, 154, 179, 213.

20 Chit Chung Ong, *Operation Matador: Britain's War Plans against the Japanese* (Singapore: Times Academic Press, 1997) is a good treatment.

21 Bennett, *Why Singapore Fell*, pp. 62–3.

22 Farrell, in Murfett *et al.*, *Between Two Oceans*, pp. 206–7. Simpson, 'Percival', p. 266.

23 Janet Uhr, *Against the Sun: The AIF in Malaya, 1941–42* (Sydney: Allen & Unwin, 1998), p. 58.

24 Lodge, *Fall of General Gordon Bennett*, pp. 45–9, and *passim*.

25 Bennett, *Why Singapore Fell*, pp. 43, 225.

26 Farrell, in Murfett *et al.*, *Between Two Oceans*, pp. 204–5.

27 Bennett, *Why Singapore Fell*, pp. 225–6.

28 Ibid., p. 47.

29 I. M. Stewart, *History of the Argyll and Sutherland Highlanders 2nd Battalion: The Malayan Campaign 1941–42* (London: Nelson, 1947). For another point of view, see Uhr, *Against the Sun*, pp. 51–3.

30 John Moremon, 'Most Deadly Jungle Fighters? The Australian Infantry in Malaya and Papua, 1941–3', BA Hons thesis, History Department, University of New England, Armidale, NSW, Australia, 1991. See also Lodge, *Fall of General Gordon Bennett*, pp. 191, 213–15.

31 That is, American, British, Dutch and Australian.

32 Lodge, *Fall of General Gordon Bennett*, pp. 84–7.

33 L. Wigmore, *The Japanese Thrust* (Canberra: Australian War Memorial, 1957), p. 214.

34 Uhr, *Against the Sun*, pp. 76–7.

35 Lodge, *Fall of the General Gordon Bennett*, ch. v, is the best analysis.

36 A. E. Percival, *The War in Malaya* (London: Eyre & Spottiswoode, 1949), p. 272.

37 Norman F. Dixon, *On the Psychology of Military Incompetence* (London: Jonathan Cape, 1976), pp. 138–40. For similar reasons the Australians were forbidden for some days to fire upon the tower of the Sultan of Johore's palace directly opposite their positions even though it was obvious the Japanese were using it as an observation post.

38 Bennett, *Why Singapore Fell*, p. 173.

39 Ibid., ch. xxii, esp. pp. 175–6; Lodge, *Fall of General Gordon Bennett*, ch. 7; and Farrell, in Murfett *et al.*, *Between Two Oceans*, pp. 227–32.

40 Percival, *War in Malaya*, p. 275.

41 Lodge, *Fall of General Gordon Bennett*, p. 154.

42 Ibid., chs 10–13. See also Mark Clisby, *Guilty or Innocent? The Gordon Bennett Case* (Sydney: Allen & Unwin, 1992).

43 Farrell, in Murfett *et al.*, *Between Two Oceans*, p. 218.

44 For a discussion of the supply situation and the other possible outcomes see my 'Churchill, Australia and the Fall of Singapore: "Inexcusable Betrayal" or Calculated Risk?', 'Churchill and Australia' conference, Menzies Centre for Australian Studies, King's College London, and Churchill Archives Centre, Cambridge, June 2002, unpub.

6 GENERAL YAMASHITA AND HIS STYLE OF LEADERSHIP: THE MALAYA/SINGAPORE CAMPAIGN

1 Arthur Swinson, *Four Samurai: A Quartet of Japanese Army Commanders in the Second World War* (London: Hutchinson, 1968), pp. 90–1.

2 Tsuji Masanobu, *Singaporu* [Singapore] (Tokyo: Tozainanboku-sha, 1952), p. 37.

3 Kunitake Teruto, 'Mare gunshireibu: Dai 25 gun kaku tatakaeri' [Headquarters of the Army in Malaya: How the 25th Army Fought], *Maru (bessatsu): Sensho no hibi*, 8 (March 1988), pp. 469–70.

4 Yasuoka Masataka, *Ningen shogun Yamashita Tomoyuki* [Humane General: Yamashita Tomoyuki] (Tokyo: Kojin-sha, 2000), pp. 354–5.

5 Oki Shuji, *Yamashita Tomoyuki* (privately published, 1958), pp. 185–6.

6 Ibid., pp. 159–60.

7 A. J. Barker, '*Mare no Tora*' *Yamashita Tomoyuki: Eiko no Singaporu koryakusen* ['Tiger of Malaya' Yamashita Tomoyuki: Glorious Capture of Singapore] (originally published by Ballantine Books in 1973 with the title *Yamashita*), trans. Toriyama Hiroshi (Tokyo: Sankei shinbun-sha shuppan-kyoku, 1976), p. 52.

8 Oki, *Yamashita Tomoyuki*, p. 360.

9 Responding to Hara Yoshimichi, Presid§ent of the Privy Council, Sugiyama explained that it would take 100 days to carry out the Malaya campaign. Boeicho

Boeikenshusho Senshishitsu [War History Office, National Defense College, Defense Agency, Japan], *Senshi-sosho: Mare shinko sakusen* [Series of Military History of World War II: Malaya Campaign] (Tokyo: Asagumo shinbun-sha, 1966), p. 113.

10 Kunitake, 'Mare gunshireibu', p. 476.

11 Boeikenshusho Senshishitsu, *Senshi-sosho: Mare shinko sakusen*, p. 167.

12 Kunitake, 'Mare gunshireibu', p. 466.

13 Boeikenshusho Senshishitsu, *Senshi-sosho: Mare shinko sakusen*, p. 167.

14 Rikusenshi kenkyu fukyu-kai (ed.), *Mare sakusen* [Malaya Campaign] (Tokyo: Hara shobo, 1966), pp. 32–4.

15 'Yamashita Tomoyuki Taisho go-shinko soan' [Draft of the Report to the Emperor by General Yamashita Tomoyuki] (Library of the National Institute for Defense Studies (hereafter as NIDS), Japan).

16 Oki, *Yamashita Tomoyuki*, p. 339.

17 Ibid., p. 180.

18 See Barker, *'Mare no Tora' Yamashita Tomoyuki*, p. 84.

19 Ibid., p. 329. On 20 November 1941, Yamashita told the officers and soldiers of the 25th Army that 'your enemies are Englishmen. The enemy's biggest disadvantage is inconsistency of race. Its spiritual solidarity is extremely weak. Strike at the English cadres and the English key units as the core of the enemy's resistance and the others will naturally collapse . . .' 'Hohei dai 11 rentai dai 3 chutai jinchu nisshi' [Staff Diary of the 3rd Company, 11th Infantry Regiment] (vol. III) (Library of NIDS).

20 *Tosuikoryo* was compiled as a guide for senior commanders who were expected to formulate plans of operations and as a book of strategic principles systematising command theories of Japanese origin.

21 Boei kyoiku kenkyukai (ed.), *Tosuikoryo, Tosuisanko* (Tokyo: Tanaka shoten, 1983), pp. 8–9.

22 A 24 howitzer regiment, a mountain artillery regiment and two trench mortar battalions were transferred to the 14th Army in the Philippines. Rikusennshi kenkyu fukyu-kai (ed.), *Mare sakusen*, p. 193.

23 Kunitake, 'Mare gunshireibu', p. 484.

24 See ibid. and Kunitake Teruto, 'Mare senki' [War Record of the Malaya/Singapore Campaign] (Library of NIDS).

25 Kunitake, 'Mare gunshireibu', p. 484.

26 Asaeda Shigeharu, 'Eikoku shijo saidai no kofuku: Sakusen sanbo no Mare senki' [The Largest Capitulation in British History: War Record of a Staff Officer in Charge of Operations], *Rekishi to jinbutsu*, Zokan: *Taiheiyosenso kaisen hiwa* (January 1983), p. 125.

27 According to S. Woodburn Kirby, the number was 88,600. S. Woodburn Kirby, *The War against Japan*, vol. I: *The Loss of Singapore* (London: Her Majesty's Stationery Office, 1957), p. 163.

28 Tsuji, *Singaporu*, p. 40.

29 Oki, *Yamashita Tomoyuki*, p. 449.

30 The 56th Division was actually committed in Burma.

31 Terazaki Ryuji, 'Ozawa Nankenkantai to Nanpogun' [Ozawa South Dispatched Fleet and the Southern Army], *Maru (bessatsu): Sensho no hibi*, 8 (March 1988), p. 105.

32 Separately from Mutaguchi's main force, Koba Task Force landed on Khota Baru and pursued Takumi Task Force. Instead of the latter, Koba Task Force conquered Endau and Mersing, the British defence positions that the Southern Army devised for attack in Plan S.

33 It seemed to be the Southern Army's initiative that Takumi Task Force should drive south along the coast to join the attack against Kuantan. Kunitake, 'Mare gunshireibu', p. 482.

34 Ibid., p. 483; Kunitake, 'Mare senki'.

35 These trucks were captured during the campaign.

36 For Mutaguchi's feelings, see Mutaguchi Renya, 'Singaporu yosai koryaku sakusen ni kansuru shoken' [Comments on the Operations in Singapore] (Library of NIDS).

37 In Takumi's case, Yamashita's warmhearted personality worked as well. See Takumi Hiroshi, *Kotabaru tekizen joriku* [Opposed Landings on Kota Bharu] (Tokyo: Puresu Tokyo, 1968), pp. 136–7.

38 Tsuji, *Singaporu*, pp. 34–5. Tsuji's criticism of the Imperial Guards Division is also supported by the division's behaviour on crossing the Johore Strait, as will be touched upon later.

39 Kunitake, 'Mare gunshireibu', pp. 479–80.

40 Yasuoka, *Ningen shogun Yamashita Tomoyuki*, p. 310.

41 Tsuji, *Singaporu*, pp. 220–2. Tsuji often went back and forth between the Command and the front lines without Yamashita's instructions, but Yamashita overlooked Tsuji's instinctive behaviour.

42 Boeikenshusho Senshishitsu, *Senshi-sosho: Mare shinko sakusen*, p. 310.

43 Kunitake, 'Mare gunshireibu', p. 477.

44 Tsuji, *Singaporu*, p. 269. During the Malaya campaign, Yamashita rested the main force of the 5th Division for two days after the occupation of Kuala Lumpur. Ibid., pp. 219–20.

45 Yasuoka, *Ningen shogun Yamashita Tomoyuki*, p. 284.

46 Boeikenshusho Senshishitsu, *Senshi-sosho: Mare shinko sakusen*, pp. 506, 565; Rikusenshi kenkyu fukyu-kai (ed.), *Mare sakusen*, p. 204.

47 'Oil tactics', igniting oil on the surface of the water, had been so feared by Yamashita that an exercise for study had been done before. Rikusenshi kenkyu fukyu-kai (ed.), *Mare sakusen*, p. 194; Yasuoka, *Ningen shogun Yamashita Tomoyuki*, p. 319.

48 Boeikenshusho Senshishitsu, *Senshi-sosho: Mare shinko sakusen*, p. 601; Rikusenshi kenkyu fukyu-kai (ed.), *Mare sakusen*, p. 229; Oki, *Yamashita Tomoyuki*, p. 381.

49 Yamashita, 'Yamashita Tomoyuki Taisho go-shinko soan'.

50 Boeikenshusho Senshishitsu, *Senshi-sosho: Mare shinko sakusen*, p. 618; Rikusenshi kenkyu fukyu-kai (ed.), *Mare sakusen*, p. 256.

51 Terazaki, 'Ozawa Nankenkantai to Nanpogun', p. 97.

52 Kunitake, 'Mare gunshireibu', p. 482.

53 Boeikenshusho Senshishitsu, *Senshi-sosho: Mare shinko sakusen*, p. 328; Endo Saburo, *Nitchu jugo-nen senso to watashi: Kokuzoku, aka no shogun to hito ha iu* [The Sino-Japanese Fifteen Years War and Me: People Call Me Traitor to the Country, Communist General] (Tokyo: Nitchu shorin, 1974), p. 220.

54 The air forces offered 60 fighters, 60 light bombers and 60 heavy bombers on the first day and 40, 40 and 60 respectively from the following day on. Kunitake, 'Mare senki'.

55 Yamashita's Army was under the orders of the Southern Army commanded by Marshal Terauchi Hisaichi. At that time Terauchi's general headquarters was located in Saigon.

56 Boeikenshusho Senshishitsu, *Senshi-sosho: Mare shinko sakusen*, p. 286.

57 Tsuji, *Singaporu*, p. 232.

58 Kunitake, 'Mare gunshireibu', pp. 481–2, 484.

59 Boeikenshusho Senshishitsu, *Senshi-sosho: Mare shinko sakusen*, p. 601; Rikusenshi kenkyu fukyu-kai (ed.), *Mare sakusen*, pp. 247–8. The vexed Command of the 25th Army dropped a communication cylinder from a friendly plane to give an order directly to Sawamura Task Force, advancing as a pursuit unit of the Infantry Corps of the Imperial Guards, bypassing its command. Rikusenshi kenkyu fukyu-kai (ed.), *Mare sakusen*, p. 254.

60 Rikusenshi kenkyu fukyu-kai (ed.), *Mare sakusen*, p. 249.

61 Ibid., pp. 248–50, 253–4.

62 Yamashita, 'Yamashita Tomoyuki Taisho go-shinko soan'.

7 BRITISH TACTICAL COMMAND AND LEADERSHIP IN THE BURMA CAMPAIGN, 1941–45

1 Arthur Swinson, *The Battle of Kohima* (London: Cassell, 1996), pp. 180–1.

2 Ibid., pp. 223–9. S. Woodburn Kirby (ed.), *History of the Second World War: The War against Japan* (vol. III) (Her Majesty's Stationery Office, 1961), p. 352. Bisheshwar Prasad (ed.), *Indian Armed Forces in World War II: Reconquest of Burma* (vol. I) (Delhi: Combined Inter-Services Historical Section (India and Pakistan), 1958), p. 298.

3 Woodburn Kirby, *The War against Japan* (vol. II), ch. 1.

4 Michael Hickey, *The Unforgettable Army* (Staplehurst: Spellmount, 1992), pp. 26–7.

5 Woodburn Kirby, *The War against Japan* (vol. II), appendix 2. Sir John Smyth, *Milestones: A Memoir* (London: Sidgwick & Jackson, 1979), pp. 152–4. Hickey, *The Unforgettable Army*, pp. 55, 62.

6 Public Record Office, London, War Office Papers [hereafter PRO WO] 172/475, 17th Indian Division G Branch War Diary.

7 Sir William Slim, *Defeat into Victory* (London: Cassell, 1956) pp. 118–19.

8 Woodburn Kirby, *The War against Japan* (vol. II), pp. 218–20. Ian Lyall Grant and Kazuo Tamayama, *Burma 1942: The Japanese Invasion* (Chichester: Zampi, 1999), ch. 25. Slim, *Defeat into Victory*, pp. 119–21.

9 Slim, *Defeat into Victory*, p. 109.

10 British Library, London, India Office Collection [hereafter BL IOC], L/WS/1/706, Burma Operations 1942–1946, Report by Sir Reginald Dorman-Smith, Governor of Burma, on the Burma Campaign 1942.

11 Slim, *Defeat into Victory*, pp. 111–12.

12 General Alexander's appendix to General Wavell's dispatch of 14 July 1942, supplement to the *London Gazette*, 11 March 1948.

13 See Slim, *Defeat into Victory*, pp. 182–96 for an explanation of his thinking on the subject.

14 PRO WO 203/5716, 17th Indian Division Report: Lessons from the Burma Campaign 1942 [The Cameron Report].

15 PRO WO 172/475, 17th Indian Division G Staff War Diary, Training Instruction No. 1, 4 June 1942.

16 Prasad, *Reconquest of Burma* (vol. I), chs 3 and 4.

17 BL IOC L/WS/1/707, Indian Army Morale and Possible Reduction, Minute for the War Cabinet Chiefs of Staff Committee, 10 May 1943, covering a Paper by the Secretary of State for India, 3 May 1943. Philip Mason, *A Matter of Honour* (London: Purnell, 1974), chs 18–20. S. P. Cohen, *The Indian Army* (Delhi: Oxford University Press 1990), chs 4–6.

18 Hickey, *The Unforgettable Army*, p. 75.

19 Woodburn Kirby, *The War against Japan* (vol. II), chs 15, 19, 20. Slim, *Defeat into Victory*, ch. 8.

20 PRO WO 172/1838, XV Corps GS War Diary 1943, Appreciation by Commander XV Corps, 26 April 1943.

21 Louis Allen, *Burma, The Longest War* (London: J. M. Dent, 1984), pp. 91–116.

22 Minute from Churchill to General Ismay for the Chiefs of Staff Committee, 8 April 1943, in Sir Winston Churchill, *The Second World War*, Vol. IV: *The Hinge of Fate* (London: Cassell, 1951), p. 841.

23 Imperial War Museum, London [hereafter IWM], Japanese AL Box 9–5030, Report by Lieutenant-General Mutaguchi Renya and Lieutenant-Colonel Fujiwara Iwaichi: The Effect on the Japanese Army of the Wingate Invasion of Spring 1943.

24 Woodburn Kirby, *The War against Japan* (vo. II), pp. 327–9. Slim, *Defeat into Victory*, pp. 162–3.

25 Slim, *Defeat into Victory*, p. 536.

26 Liddell Hart Centre for Military Archives, King's College, London [hereafter LHCMA], Messervy Papers 5/1–13, General Officer Commanding 7th Indian Division Operational Notes. National Army Museum, London [hereafter NAM], Army in India Training Memorandum No. 24, March 1944. The text is lifted straight from Messervy's Operational Notes.

27 Prasad, *Reconquest of Burma* (vol. I), ch. 10.

28 PRO WO 203/2468, 17th Division Operations on the Tiddim Road, Report by the US Military Attaché in India,14 December 1943.

29 PRO WO 203/682 and 203/684 give a detailed analysis of the Kyaukchaw battle on the 20th Division front on 10 February 1944.

30 BL IOC L/WS/1/707, Indian Army Morale and Possible Reductions, Telegram from Viceroy to Secretary of State for India, 29 July 1943.

31 NAM, Army in India Training Memorandum No. 23, December 1943.

32 PRO WO 106/4708, Jungle Warfare Training, Report by the Infantry Committee MO12,11 October 1943.

33 BL IOC L/WS/1/707, Indian Army Morale and Possible Reductions, Letter from the Chief of the General Staff (India) to the Chief of the Imperial General Staff, 9 September 1943; Minute for the War Cabinet Chiefs of Staff Committee, 10 May 1943. BL IOC L/WS/1/939, Quarterly Reports of Adjutant-General (India) Committee on Morale, Report No. 4 August–October 1943.

34 Cohen, *The Indian Army*, p. 145.

35 Slim, *Defeat into Victory*, p. 166.

36 BL IOC L/WS/1/441, Liaison Letters from the Chief of the General Staff (India) to the Chief of the Imperial General Staff, Letter 16576/II/1/SD1,15 February 1944.

37 Hickey, *The Unforgettable Army*, p. 184.

38 BL IOC L/WS/1/441, Liaison Letters from the Chief of the General Staff (India) to the Chief of the Imperial General Staff. BL IOC L/WS/1/939, Quarterly Reports of the Adjutant-General (India) Committee on Morale. PRO WO 203/5203, Morale in South-East Asia Command July 1944 to June 1945. PRO WO 203/ 4536, South-East Asia Command Morale Reports June to November 1944. PRO WO 203/4538, Allied Land Forces South-East Asia Morale Reports August to November 1944. PRO WO 193/452–453, Morale Committee Agenda, Minutes and Papers.

39 Woodburn Kirby, *The War against Japan* (vol. III), ch. 10.

40 Slim, *Defeat into Victory*, p. 246.

41 The British fielded the 5th, 7th, 26th, 36th and 81st Divisions against the Japanese 55th Division. The Allies also had overwhelming air superiority by this time.

42 Swinson, *The Battle of Kohima*. C. E. Lucas Phillips, *Springboard to Victory* (London: Heinemann, 1966).

43 BL IOC L/WS/1/777, Command Study Period: The Burma Campaign, November 1944, Lecture by Brigadier I. M. Stewart. NAM, Army in India Training Memorandum No. 25, July 1944.

44 Swinson, *The Battle of Kohima*, pp. 138–9.

45 Ian Lyall Grant, *Burma – The Turning Point* (Chichester: Zampi, 1993), ch. 10. IWM, Japanese AL Box 21–5240, The British Road Block on the Tiddim Road. IWM, Japanese AL Box 18–5202, Burma Operations Record: 15th Army Operations in the Imphal Area, Japanese Monograph 134, p. 65.

46 John Leyin, *Tell Them of Us* (Stanford le Hope, Essex: Leyins, 2000), pp. 192–3. Woodburn Kirby, *The War against Japan* (vol. III), pp. 133–52.

47 PRO WO 203/1897, Infantry/Operations Reports. See also IWM Japanese AL Box 24–5295, Report of 55th Division Operations in Arakan February 1944, which mentions the effect of British combined arms tactics.

48 PRO WO 172/4247, 2nd British Division G Branch War Diary, Operation Instruction No 11: Clearance of Kohima Ridge. Woodburn Kirby, *The War against Japan* (vol. III), chs 13, 14. Prasad, *Reconquest of Burma* (vol. I), ch. 15. Swinson, *The Battle of Kohima*, chs 8–11.

49 Swinson, *The Battle of Kohima*, pp. 170–1.

50 Prasad, *Reconquest of Burma* (vol. I), ch. 4.

51 Slim, *Defeat into Victory*, p. 357.

52 Ibid., pp. 541–2.

53 Ibid., pp. 294–5.

54 Ibid., pp. 309–10.

55 PRO WO 203/1239, XIV Army Planning Instruction No. 6, 17 January 1945.

56 Slim, *Defeat into Victory*, pp. 521–2. See also Morale Reports in PRO WO 203/5203, Morale in South-East Asia Command July 1944 to June 1945.

8 JAPANESE WAR LEADERSHIP IN THE BURMA THEATRE: THE IMPHAL OPERATION

1 S. Woodburn Kirby, *The War against Japan*, (London: Her Majesty's Stationery Office, 1958), p. 1.

2 Boeikenshusho Senshishitsu [War History Office, National Defense College, Defense Agency, Japan], *Senshi-sosho: Burma Kouryaku Sakusen* [Series of Military History of World War II: The Taking of Burma] (Tokyo: Asagumo Shinbun-sha, 1968), p. 91.

3 Mutaguchi Renya, 'U-go Sakusen ni kansuru kokkai Toshokan ni okeru setsumei siryou', (Explanatory papers about Operation U-go: Imphal operation), National Diet Library, Japan.

4 This paper has benefited from many preceding studies. I would especially like to mention Tobe Ryoichi, *Shippai no Honsitsu* [Nature of Failure] (Tokyo: Daiyamond-sha, 1984), pp. 92–119. Isobe Takuo, *Inparu Sakusen* [Operation Imphal] (Tokyo: Marunouchi Shuppan, 1984). My paper is only complementary to these studies.

5 *Senshi-sosho: Burma Kouryaku Sakusen*, pp. 562–3.

6 Ibid., p. 565.

7 In October 1942, Major-General Isayama was dismissed as Chief of Staff of 15th Army, for he mentioned openly Iida's opinion about Operation 21 at the conference in IGHQ.

8 Mutaguchi Renya, 'Inparu Sakusen Kaisoroku 1' [Memoir of Operation Imphal 1] (1956), pp. 56–7, Library of the National Institute for Defense Studies (hereafter as NIDS), Japan.

9 Boeikenshusho Senshishitsu, *Senshi-sosho: Inparu Sakusen* [Series of Military History of World War II: Operation Imphal] (Tokyo: Asagumo Shinbun-sha, 1968), pp. 90–1.

10 Ibid., pp. 547–8.

11 Defence by means of an offensive operation (a pre-emptive attack) assumes an impending total offensive by the enemy. No offensive is necessary if no enemy attack is anticipated.

12 *Senshi-sosho: Inparu Sakusen*, p. 97.

13 15th Division was not yet incorporated into the Burma Theatre Army at this time.

14 *Senshi-sosho: Inparu Sakusen*, pp. 112–13.

15 Ibid., p. 119. IGHQ redeployed two divisions from the Pacific theatre to the Burma theatre.

16 Katakura Tadashi, *Inparu Sakusen hishi* [Secret History of the Imphal Operation] (Tokyo: Keizaiorai-sha, 1975), p. 103.

17 Ibid., p. 104.

18 He was transferred to 19th Army HQ.

19 Hattori Takushiro, *Daitoa Senso Zenshi* [Complete History of the Great East Asian War] (Tokyo: Hara Shobo, 1968), p. 594. It is important to note that the IGHQ issued a directive, not an order. Orders required the Emperor's signature, whereas directives did not.

20 Isobe Takuo, *Inparu Sakusen* [The Imphal Operation: Experience and Studies] (Marunouchi Shuppan, 1984), p. 196; Louis Mountbatten, 'South East Asia 1943–45: Report to the CCS'.

21 John Ehrman, *Grand Strategy* (London: Her Majesty's Stationery Office, 1956), p. 148.

22 Viscount Slim, *Defeat into Victory* (London: Cassell, 1956), pp. 291–2.

23 On 8 December 1941, when Subhas Chandra Bose heard that Japan had declared war against the United States and Britain, he was in Berlin. He requested permission from the Japanese ambassador to go to Japan, but the IGHQ did not provide its approval until August 1942 and he did not arrive until May 1943.

24 Yomiuri Shinbun-sha (ed.), *Showa-shi no Tenno*, 9 [The Emperor in the History of the Showa period] (Tokyo: Yomiuri Shinbun-sha, 1969), p. 34.

25 John Ferris, 'Warera jishin ga eranda senjo' [The Battlefield We Chose by Ourselves], in Hirama Yoichi, Ian Gow and Hatano Sumio (eds), *Nichi ei koryu shi, 1600–2000*, Vol. III: *Gunji* [The History of Anglo-Japanese Relations, 1600–2000, Vol. III: The Military Dimension] (Tokyo: Tokyo Daigaku Shuppankai, 2001), p. 226.

26 'Just before we could launch the Imphal operation, Wingate's airborne troops began to land in our area. So the discussion grew heated as to whether the Imphal operation should start or not. At that time I wrote a private letter to General Tojo to request as many reinforcement troops with anti-aircraft guns as possible and glider airborne forces for the 15th Army. In that way, I would be able to maintain our supply organisation' Mutaguchi, 'Inparu Sakusen Kaisoroku 1'.

27 Reportedly, the headquarters were located distant from the theatre in order to cope with enemy airborne attacks.

28 Koseisho Fukuinkyoku [Demobilization Department, Ministry of Health and Welfare] (co-ordinated), 'Biruma Sakusen Kiroku: Inparu Homen Dai 15 gun no Sakusen' [Records of the Burma Operation: Operations of the 15th Army in Imphal] (1949), p. 161 Library of NIDS.

29 Hohei Dai 215 Rentai Senki Hensan iinkai, [Editing Committee for the War Records of the 215th Infantry Regiment], *Hohei Dai 215 Rentai Senki* [War Records of the 215th Infantry Regiment] privately published edition, 1972, pp. 538–58. According to this record, Imphal survivors from the 1st and 3rd Battalions of the 215th Infantry Regiment in charge of cutting off the enemy retreat said, 'We could have somehow held one position out of three. But we could not have held the other two positions against a large enemy spillover and break-through.'

30 *Senshi-sosho: Inparu Sakusen*, p. 460.

31 Mutaguchi, 'U-go Sakusen ni kansuru kokkai Toshokan ni okeru setsumei siryou', A. J. Barker's letter.

32 Arthur Swinson, *Kohima* (London: Cassell, 1966), p. 65. Translated into Japanese by Nagao Mutsuya, *Kohima* (Tokyo: Hayakawa Shobo, 1967), p. 81.

33 *Senshi-sosho: Inparu Sakusen*, pp. 498–502. S. Woodburn Kirby, *The War against Japan* (London: Her Majesty's Stationery Office, 1961), p. 300.

34 *Senshi-sosho: Inparu Sakusen*, p. 502.

35 Takagi Toshio, *Funshi* [Death in Anger] (Tokyo: Bungeishunju-sha, 1969), pp. 13–16.

36 Ushiro Masaru, *Biruma Senki* [The War in Burma] (Tokyo: Nihon shuppan kyodo kabushiki gaisha, 1953), p. 30.

37 *Senshi-sosho: Inparu Sakusen*, pp. 560–2.

38 Ibid., p. 570.

39 Hohei Dai 215 Rentai Senki Hensan iinkai, *Hohei Dai 215 Rentai Senki*, p. 590.

40 Ibid., pp. 581–7.

41 *Senshi-sosho: Inparu Sakusen*, pp. 542, 586. Rikusenshi kenkyu fukyu-kai, *Rikusenshishu* [Series of Ground Warfare History], Vol. XVII (Tokyo: Hara Shobo, 1970), pp. 202–6.

42 *Senshi-sosho: Inparu Sakusen*, p. 612.

43 Ibid., pp. 611–12.

44 Boeikenshusho Senshishitsu, *Senshi-sosho: Irawaji Kaisen* [Series of Military History of World War II: The Irrawaddy Crossing] (Tokyo: Asagumo Shinbun-sha, 1969), pp. 63–4.

45 Yamamoto Tsunetomo, *Edo shiryo sosho: Hagakure (Jo)* [Series of source books of the Edo periods: Hagakure (Hiding behind the Leaves), Vol. I] (Tokyo: Jinbutsuorai-sha, 1968), p. 100. Yamamoto Tsunetomo was a samurai who served under Clan Nabeshima in what is now presently Saga prefecture, Kyushu in the seventeenth-century Edo period. General Mutaguchi was born and brought up in Saga prefecture.

9 BRITISH LEADERSHIP IN AIR OPERATIONS: MALAYA AND BURMA

1 Philip Ziegler, *Mountbatten: The Official Biography* (Glasgow: Fontana/Collins,1986) p. 225.

2 Henry Probert, *The Forgotten Air Force: The Royal Air Force in the War against Japan, 1941–1945* (London: Brassey's,1995), p. 25.

3 Ibid., pp. 21–3.

4 Ibid., pp. 28 ff.

5 Ibid., pp. 55–63.

6 Ibid., p. 84.

7 Ibid., pp. 84–92.

8 Raymond Callahan, *Burma 1942–1945* (London: Davis-Poynter, 1978), pp. 68–106.

9 Ziegler, *Mountbatten*, pp. 186–96, 216–26.

10 For the first Chindit operation see S. Woodburn Kirby, *The War against Japan*, vol. II: *India's Most Dangerous Hour* (London: Her Majesty's Stationery Office, 1958), pp. 309–29, and also Louis Allen, *Burma: The Longest War, 1941–45* (London: Phoenix, 1988), pp. 116–49. Callahan, *Burma*, p. 79.

11 Brian Bond (ed.), *Chief of Staff: The Diaries of Lieutenant-General Sir Henry Pownall* (2 vols) (London: Leo Cooper, 1974), vol. II *1940–1944*, Diary entries, 17 October 1943 and 19 February 1944, pp. 112, 142.

12 Allen, *Burma*, p. 148.

13 Callahan, *Burma*, p. 79.

14 Ronald Lewin, *Slim: The Standard Bearer: A Biography of Field-Marshal The Viscount Slim* (London: Leo Cooper, 1976), p. 111.

15 Probert, *Forgotten Air Force*, pp. 106–7.

16 Ibid., p. 106.

17 Pownall diaries, vol. II, entry 28 October 1943, p. 117.

18 Probert, *Forgotten Air Force*, pp. 222, 223.

19 Ibid., p. 222.
20 Ibid., p. 224.
21 Ibid., p. 223.
22 Ibid., p. 122–3.
23 Ibid., p. 107.
24 Ibid., pp. 107–20.
25 Woodburn Kirby, *The War against Japan* (vol. III), p. 386.
26 Probert, *Forgotten Air Force*, p. 120.
27 Pownall diaries, vol. II, entry 22 November 1943, p. 119.
28 S. Woodburn Kirby, *The War against Japan*, vol. III: *The Decisive Battles* (London: Her Majesty's Stationery Office, 1962), pp. 47–8. Probert, *Forgotten Air Force*, pp. 149–50. Philip Ziegler (ed.), *Personal Diary of Admiral the Lord Louis Mountbatten: Supreme Allied Commander, South East Asia, 1943–1946* (London: Collins, 1988), entry 11 December 1943, pp. 40–1.
29 Probert, *Forgotten Air Force*, p. 150.
30 Woodburn Kirby, *The War against Japan* (vol. III), pp. 122, 205. Probert, *Forgotten Air Force*, p. 162.
31 Kirby, *The War against Japan* (vol. III), pp. 127–8. Lewin, *Slim*, pp. 151, 157. Probert, *Forgotten Air Force*, p. 168.
32 For details of the Imphal and Kohima campaigns see Kirby, *The War against Japan* (vol. III), chs XVI, XXI, XXIII–XXV. For Allied air activity see Probert, *Forgotten Air Force*, pp. 181 ff.
33 Probert, *Forgotten Air Force*, p. 184. Kirby, *The War Against Japan* (vol. III), pp. 450–1, Lewin, *Slim*, pp. 174–5. Pownall diaries, vol. II, entry 18 March 1944, pp. 141–2.
34 Lewin, *Slim*, p. 175.
35 Ibid., p. 175.
36 Woodburn Kirby, *The War against Japan* (vol. III), pp. 325–7.
37 Probert, *Forgotten Air Force*, p. 188.
38 Ibid., p. 186.
39 Woodburn Kirby, *The War against Japan* (vol. III), pp. 387–9.
40 Probert, *Forgotten Air Force*, p. 171.
41 Lewin, *Slim*, p. 143.
42 Ibid., p. 144.
43 Callahan, *Burma*, pp. 138–9; Pownall diaries, entry 1 April 1944, p. 157.
44 Probert, *Forgotten Air Force*, pp. 246–9. For the land operations see Woodburn Kirby, *The War against Japan*, vol. IV: *The Reconquest of Burma* (London: Her Majesty's Stationery Office, 1969), chs XXIII–XXXII.
45 Lewin, *Slim*, p. 158.
46 Woodburn Kirby, *The War against Japan* (vol. IV), pp. 409–12.
47 Lewin, *Slim*, p. 175.
48 Pownall diaries, entry 14 September 1943, p. 108.
49 Ibid., entry 22 November 1943, p. 118.
50 Probert, *Forgotten Air Force*, pp. 302–3.
51 Ibid., p. 225.
52 Ibid., p. 245.
53 Ibid., p. 131.
54 Ibid., p. 168.

55 Pownall diaries, entry 9 April 1944, p. 160.
56 Woodburn Kirby, *The War against Japan* (vol. III), p. 373–4.

10 AIR OPERATIONAL LEADERSHIP IN THE SOUTHERN FRONT: IMPERIAL ARMY AVIATION'S TRIAL TO BE AN 'AIR FORCE' IN THE MALAYA OFFENSIVE AIR OPERATION

1 Boeicho Boeikenshusho Senshishitsu [War History Office, National Defense College, Defense Agency, Japan], *Senshi-sosho: Rikugunkouku no Gunbi to Unyo (2)* [Series of Military History of World War II: Armaments and Operations of Imperial Army Aviation (2)] (Tokyo: Asagumo Shinbun-sha, 1974), pp. 560–1.
2 Denis Richards and Hilary St George Saunders, *Royal Air Force 1939–1945, Volume II* (London: Her Majesty's Stationery Office, 1954), p. 23.
3 Rikugun Kouku-honbu, 'Kouku: butai Yohou (fukumu Dou Hensanriyusho)' [The Army Air Service Head Office, 'Air Troops Operations: Including the Reason for compilation'] (1937), Library of the National Institute for Defense Studies, Defense Agency, Japan (hereafter as NIDS).
4 Koseisho, Dai 1 Fukuinnkyoku [The First Demobilisation Department, Ministry of Health and Welfare] (ed.), 'Rikugun Kouku Ennkakushi: Kouku-yohei' [The History of Imperial Army Aviation: Air Operations] Library of NIDS.
5 Hisayuki Yokoyma, 'Military Technological Strategy and Armaments Concepts of the Japanese Imperial Army: Around the Post-WWI Period', *NIDS Security Reports*, No. 2 (March 2001).
6 'Rinji Kouku-jutu Rensyu-iin Kankeishorui: Futu-koku Koukudan Kannkei' [The Training Report on Faure's Wing] Library of NIDS.
7 'Koukuhei Souten Seitei no ken' [Imperial Army's diary in 1934 to file establishment document of 'Airman Drill Book'], Ministry of the Army, *Dai-nikki Koji Shouwa 9*, Library of NIDS.
8 Boeikenshusho Senshishitsu, *Senshi-sosho: Rikugunkouku no Gunbi to Unyou (1)* [War History Series: Armaments and Operations of Imperial Army Aviation (1)] (Tokyo: Asagumo Shinbun-sha, 1971), p. 433.
9 Ibid., p. 439.
10 Rikugun Kouku-honbu, 'Kouku: butai Yohou (fukumu Dou Hensanriyusho)'.
11 *Senshi-sosho: Rikugunkouku no Gunbi to Unyou (1)*, p. 528.
12 Ibid., pp. 517–18.
13 'Ohshima Kendoku Kouku-shisatsudan Houkoku Dai-ikkan' [The Report of the Ohshima Air System Inspection Delegation to Germany 1936, Vol. I], Library of NIDS.
14 A border incident between Japan and the Soviet Union, this incident broke out at the Khalkha River on the Mongolian border in 1939.
15 'Kouku Sakusen Kouyo; Showa 15' ['Air Troops Operations' in 1940], Library of NIDS.
16 Boeikenshusho Senshishitsu, *Senshi-sosho: Nanpo Shinko Rikugun Kouku-sakusen* [The Offensive Operations of Imperial Army Aviation in the Southern Area] (Tokyo: Asagumo Shinbun-sha, 1970), p. 14.
17 Ibid., pp. 28–9.

18 Koseisho, Fukuinkyoku [Demobilisation Department, Ministry of Health and Welfare] (ed.), 'Tanigawa Kazuo Shosho Zuisouroku' [Tanigawa's Diary at the Chief of the Air Staff Office at the Southern General Army Headquarters], Library of NIDS.

19 *Senshi-sosho: Nanpo Shinko Rikugun Koku-sakusen*, p. 34.

20 'Nanpo Zepan Kouku-sakusen Kiroku' [The Document of Whole Air Operations in the Southern Area], Library of NIDS.

21 The Documents of Ishii, No.35, 'Nanpogun no Sakusen-jyunbi' [Preparation for Operation at the Southern General Army], Library of NIDS. These documents were filed by Colonel Ishii Masayoshi who was an operational officer of the Southern General Army.

22 *Senshi-sosho: Nanpo Shinko Rikugun Kouku-sakusen*, pp. 182–4.

23 Ibid., pp. 71–7.

24 'Tanigawa Kazuo Shosho Zuisouroku'.

25 *Senshi-sosho: Nanpo Shinko Rikugun Kouku-sakusen*, p. 329.

26 'Tanigawa Kazuo Shosho Zuisouroku'.

27 *Senshi-sosho: Nanpo Shinko Rikugun Kouku-sakusen*, p. 114.

28 'Dai 3 Hikoushudan Jyokyo-houkoku' [The 3rd Air Corps' Report to the Commander of the Southern General Army on 17 February 1942], Library of NIDS.

29 'Kouku wo hikiite kanari: Shina Jihen Jyugun Shoken' [Sugawara's Opinion about Air Operations in the Sino-Japanese War of 1937–45 to the Minister of War in 1939], Library of NIDS.

30 'Dai 3 Hikoushudancho Sugawara Michioho Nikki' (a handwritten copy), 20 October 1941 [Sugawara's diary from 1 September 1941 to 20 May 1942], Library of NIDS.

31 'Sugawara Shougun no Nikki' [General Sugawara's Diary], *Kaiko*, 519 (Kaiko-sha, March 1994), p. 17.

32 'Dai 3 Hikoushudancho Sugawara Michioho Nikki', 20 October 1941.

33 'Ohshima Kendoku Kouku-shisatudan Houkoku Dai-ikkan'.

34 'Sugawara Michioho Chujou Kaisouroku: Ikusa no Chiri' [Sugawara's memoirs, 1954], Library of NIDS.

35 'Dai 3 Hikoushudancho Sugawara Michioho Nikki', 31 December 1941.

36 Ibid.

37 'Dai 3 Hikoushudancho Sugawara Michioho Nikki', 8 December 1941.

38 'Sugawara Michioho Chujou Kaisouroku: Ikusa no Chiri'.

39 'Dai 3 Hikoushudancho Sugawara Michioho Nikki', 1 January 1942.

40 *Senshi-sosho: Nanpo Shinko Rikugun Kouku-sakusen*, p. 352.

41 'Sugawara Shogun no Nikki', pp. 32–3.

42 'Dai 3 Hikoushudancho Sugawara Michioho Nikki', 27 December 1941.

43 Ibid.

44 'Kouku Sakusen Kouyo: Showa 15'.

45 'Rikugun Kouku Enkakushi: Kouku-yohei'.

46 'Newman Bunsho' [The document on a pulse-radar's principle that British soldier Newman had], Library of NIDS.

47 Usami Shouzo, 'Nihon no Radar no Hanashi', in Yagi Kazuko (ed.), *Radar no Shijitu* [The History of the Japanese Army's Radar Development] (privately published edition, 1995), p. 20.

48 'Nanpogun Soshireikan ni taisuru Di 3 Hikoushudan (Dai 3 Kouku) Jyokyo Houkoku' [The 3rd Air Army's Report to the Commander of the Southern General Army on 1 May 1943], Library of NIDS. The 3rd Air Corps was upgraded to 3rd Air Army in July 1942.
49 'Kouku Sakusen Kouyo: Showa 19' (1944), Library of NIDS.
50 'Kouku wo hikiite kanari: Shina Jihen Jyugun Shoken'.

INDEX